HIGHER
Maths

grade **booster**

✕ Edward Mullan ✕

ISBN 978-1-84372-483-4

Published by
Leckie & Leckie Ltd
An imprint of HarperCollins*Publishers*
Westerhill Road, Bishopbriggs, Glasgow, G64 2QT
T: 0844 576 8126   F: 0844 576 8131
leckieandleckie@harpercollins.co.uk   www.leckieandleckie.co.uk

Special thanks to
Pumpkin House (illustration), BRW (design and page makeup), Jill Laidlaw (copy-editing),
Caleb Rutherford (cover-design) and Tara Watson (proofreading).

A CIP Catalogue record for this book is available from the British Library.

® Leckie & Leckie is a registered trademark

Acknowledgements
Leckie & Leckie has made every effort to trace all copyright holders.

If any have been inadvertently overlooked, we will be pleased to make the necessary arrangements.

MIX
Paper from
responsible sources
FSC C007454

FSC™ is a non-profit international organisation established to promote the responsible management of the world's forests. Products carrying the FSC label are independently certified to assure consumers that they come from forests that are managed to meet the social, economic and ecological needs of present and future generations, and other controlled sources.

Find out more about HarperCollins and the environment at
www.harpercollins.co.uk/green

CONTENTS

1 The Structure of the Course

In common with other Higher subjects, the Maths course is divided into 3 units. This provides a flexibility for presenting centres and candidates in their management of time and resources that would be impossible with a single, integrated course.

Details of the course and other useful materials can be downloaded from the SQA Maths Website at: *http://www.sqa.org.uk/sqa/controller?p_service=Content. show&p_applic=CCC&pContentID=2464*

If you don't want to type all that in, just Google for 'SQA', select 'SQA NQ homepage' and when you get there, choose 'Mathematics'.

The syllabus is designed to build on the knowledge of the course content of either the Credit Standard Grade course or the Intermediate 2 (including Unit 3) course. For the sake of simplicity, these will be referred to as 'Credit skills'.

The course puts an emphasis on problem solving and integrating the knowledge and skills across the whole syllabus. If you are coming to Higher via the Intermediate route, you should be aware that these aspects are not, in general, a major feature of these courses

The following table summarises the Higher maths course.
Remember that the examination will try to provide a balance of marks for each topic.

	Topic 1	Topic 2	Topic 3	Topic 4
Maths 1	The straight line	Functions/graphs	Differentiation	Recurrence relations
Maths 2	Factor/Remainder Theorem; quadratic theory	Integration	Trigonometric formulae	The circle
Maths 3	Vectors	Further calculus	Logarithms and exponential functions	The wave function

Skills and knowledge from one topic may be needed when tackling a problem set in another topic.

In the SQA Conditions and Arrangements document the skills required to achieve a C-grade pass are listed in a Times font. The skills and/or performance required to achieve an A/B grade are highlighted in Arial.

The device [A/B] also appears at the end of each such reference.

Questions aimed at testing this level of skill will generally appear at the end of a paper or at the end of a question, where C-grade lead-ins may precede them.

2   The Types of Questions

Objective questions

Short response questions

Extended response questions

In the final exam there will be three types of question;

- objective questions
- short response questions
- extended response questions

Objective questions

In the Higher Maths exam, an objective question has four possible answers. Only one is correct. The others act as distracters that look sensible but are actually incorrect. These questions are used to check your breadth of knowledge of the course content and are worth two marks. No marks will be awarded for a wrong answer.

The SQA have produced guidelines for answering these questions. You'll be issued with a guide before the exam. This will give you examples of questions and instructions on how to answer them. These instructions will be repeated in the exam paper but to save time you should be familiar with the instructions before you go into the exam.

It is a good idea to number your rough working even though it will not be marked. When you come to check over your paper, the numbering will make the task easier.

If you do not have this guide, a copy can be downloaded from the SQA site. (See page 5.) Follow the link to 'Higher Objective Questions.'

Short response questions

These questions usually test an essential skill.

> **For example, from paper 1, 1998.**
>
> **Solve the equation** $2 \sin\left(2x - \dfrac{\pi}{6}\right) = 1, \ 0 \le x \le 2\pi$
>
> 4 marks

Response

This is a single question with a response that needs one strategy.

$$\sin\left(2x - \frac{\pi}{6}\right) = \frac{1}{2}$$

$$\Rightarrow 2x - \frac{\pi}{6} = \frac{\pi}{6}, \ \frac{5\pi}{6}$$

$$\Rightarrow 2x = \frac{2\pi}{6}, \ \frac{6\pi}{6}$$

$$\Rightarrow x = \frac{\pi}{6}, \ \frac{\pi}{2} \text{ and no other solutions}$$

Working in degrees will lose marks and giving more answers than are needed will lose marks.

Extended response questions

These questions generally require the use of more than one mathematical skill or area of knowledge. They are used to check your ability to undertake extended thinking, problem solving and to integrate knowledge and skills across the curriculum.

For example, from paper 2, 2001.

Triangle ABC has vertices A(2, 2), B(12, 2) and C(8, 6).

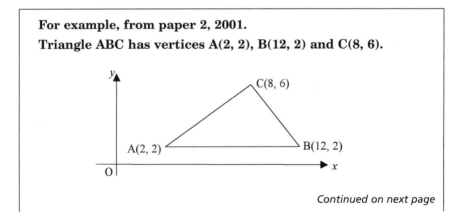

Continued on next page

> **(a)** Write down the equation of l_1, the perpendicular bisector of AB.   1 mark
>
> **(b)** Find the equation of l_2, the perpendicular bisector of AC.   4 marks
>
> **(c)** Find the point of intersection of lines l_1 and l_2.   1 mark
>
> **(d)** Hence find the equation of the circle passing through A, B and C.   2 marks

The allocation of marks should provide you with some idea of the amount of work or number of steps required in a response.

See how the question brings together the equation of a straight line, basic geometry facts from Credit and the equation of a circle.

By breaking the question into four parts the examiner helps the candidate through the overall strategy, ensuring at least an entry into the question. Since marks make grades, always make sure you attempt each question. Your initial feeling might be that a question is too hard, but make sure you pick up these early marks. They may separate a fail from a C, a C from a B or a B from an A.

The question could have been 'Find the equation of the circle passing through A(2, 2), B(12, 2) and C(8, 6)' which would have made all 8 marks inaccessible to the students who don't 'click' that the centre lies on the intersection of the perpendicular bisectors of two chords.

> (i) The perpendicular bisector of a chord of a circle passes through the centre of that circle.
>
> (ii) The centre of the circle must lie on the intersection of two such perpendicular bisectors.
>
> (iii) The radius can be found by working out the distance from this centre to any one of A, B or C.

(Response)

(a) $x = 7$ (the key words 'Write down ...' and the solitary mark hints that the answer is obvious). AB runs parallel to the x-axis so its perpendicular bisector runs through its midpoint (7, 2) parallel to the y-axis.

(b) midpoint is (5, 4)

$$m_{AC} = \frac{2}{3}$$

$$m_{\perp} = -\frac{3}{2}$$

$$y - 4 = -\frac{3}{2}(x - 5)$$

Each line is worth a mark ... so each line should be evident.

(c) Substitute $x = 7$ into $y - 4 = -\frac{3}{2}(x - 5)$ and solve for y

$$y - 4 = -3 \Rightarrow y = 1$$

point of intersection P(7, 1)

(d) radius is distance from, say A to P:

$$AP = \sqrt{[(7 - 2)^2 + (1 - 2)^2]} = \sqrt{26}$$

Equation of circle centre (7, 1) and radius $\sqrt{26}$ is

$$(x - 7)^2 + (y - 1)^2 = 26$$

3 The Structure of the Exam

Paper 1

Paper 2

Content balance

C and A/B balance

ic, ss, pd balance

Calculator/non-calculator/calculator neutral balance

The given formulae

Getting the grades

The exam is in two parts.

Paper 1

Paper 1 is 90 minutes long and has two sections, A and B.

Section A is made up of 20 multiple choice (objective test) questions each worth two marks.

Section B is worth another 30 marks and comprises approximately three to five short and extended response questions.

At least 75% of the objective test questions will be testing C-grade skills.

Calculators are not allowed in this paper.

Paper 2

Paper 2 is 70 minutes long.

It contains a mixture of short and extended response questions worth a total of 60 marks.

Although a calculator is permitted, not all the questions will need one. These questions are called calculator neutral questions.

There are between six and eight questions in this paper.

When constructing the exam the setters try to strike a balance between various factors.

Content balance

The examiners try to balance the marks evenly throughout the 12 topics. (See the table on page 5.)

At least a quarter of the marks are allocated to each of the three units.

The examiners will merge some topics in a single question.

This is unlikely to happen in the multiple choice (objective test) questions, which tend to be on a single topic.

C and A/B balance

Three-quarters of the multiple choice questions will be testing C-grade skills.

Over the whole exam approximately 65% of the marks will be for C-grade skills.

The other 35% will be testing the list of A/B skills. You can find lists of A/B-grade and C-grade skills in the appendix 3 to this book at *www.leckieandleckie.co.uk*.

ic, ss, pd balance

When marking an answer to a problem, the examiners award marks under three broad categories labelled *ic*, *ss* and *pd*.

ic stands for 'interpretation and communication' marks. Typically these are awarded when you read a question, a diagram, an equation, a formula or notation and glean data from it, (*interpretation*) or communicate the most relevant features, facts or answers (*communication*).

ss stands for 'selecting a strategy'. Strategy marks are available when the method of solving the problem is chosen, e.g. for knowing to differentiate.

pd stands for 'processing data'. The processing of data at Higher can quite often be of data in symbolic form e.g. algebraic manipulation, solving equations or performing algorithms such as integration.

Marks in the exam are balanced to provide 25% *ic*, 25% *ss* and 25% *pd* with the rest of the marks distributed any way between the three types.

More detailed lists of these features (called GRC or Grade Related Criteria) can be found in appendix 3 to this book at *www.leckieandleckie.co.uk*.

Calculator/non-calculator/calculator neutral balance

Paper 1 tests your non-calculator skills.

Paper 2 contains any question in the exam that needs a calculator.

In both papers there are calculator neutral-questions – questions that do not need a calculator.

The given formulae

Certain formulae are given for use during the exam so these do not need to be memorised. However you do need to know when to use them. Marks are often awarded at the point where substitution into the formula occurs. It is therefore a good idea to perform the complete substitution before performing any manipulations or evaluations. That way the *ss* mark is safe even if errors mean the loss of the subsequent *pd* mark.

The given formulae are listed in appendix 4 to this book at *www.leckieandleckie.co.uk*.

This book is arranged so that each of the 12 topics are covered.

A list of necessary knowledge is followed by a list of typical problems you should be able to solve. Then there are some questions which exhibit each type of multiple choice, non-calculator extended response question and calculator extended response question. For each question the allocation of marks to a correct response will be given, highlighting dos and don'ts to optimise your chance of full marks.

Getting the grades

In the mathematics exam a total of 130 marks are available.
Each mark is awarded for performing a task or exhibiting a skill.
None of the marks are for the quality of the answer so there is no particular A-grade response to a question. You can either do the question or you can't.
The examiner will differentiate between the C-grade candidate and the A-grade candidate by the marks he or she gets.
The SQA have provided the following table as a rough guide to where the cut-off scores are for the different grades.

For a total mark range of 0–130, the following (taken from an SQA specimen paper document) gives an indication of the cut-off scores:

Grade	Band	Mark Range
A	1	111–130
A	2	91–110
B	3	85–90
B	4	78–84
C	5	72–77
C	6	65–71
D	7	58–64
NA	8	52–57
NA	9	45–51

These cut-off scores may be lowered if the question papers turn out to be more demanding, or raised if the question papers are less demanding.

You can use this table and your prelim mark to help you gauge if you are on-target or not for your desired grade and to judge how much work you need to put in before the big day.

In the following chapters, each topic in each unit follows the contents checklist issued by the SQA as closely as possible. Although aimed primarily at the people who deliver the courses the SQA checklist can be of help to candidates.

Properties of the Straight Line

- *What you should know*
- *Related problems*
- *Objective questions*
- *An extended response question that doesn't need a calculator*
- *Extended response questions that need a calculator*

What you should know

You are expected to know the following facts.

- The gradient of a line passing through $A(x_a, y_a)$ and $B(x_b, y_b)$:
  $$m_{AB} = \frac{y_b - y_a}{x_b - x_a} ; x_a \neq x_b .$$

- The distance formula $AB = \sqrt{(x_b - x_a)^2 + (y_b - y_a)^2}$.

- The gradient of a line making an angle of $\theta°$ with the positive direction of the x-axis (scales being equal):
  $m = \tan\theta°$.

- A locus is a set of points fitting a particular defining description.

- The locus of the points (x, y) fitting the description $ax + by + c = 0$, where a and b are not both zero is a straight line.

- The line passing through (x_1, y_1) with gradient m has equation
  $y - y_1 = m (x - x_1)$.

- Lines are parallel if and only if their gradients are equal.

- Lines with non-zero gradients m_1 and m_2 are perpendicular if and only if $m_1 m_2 = -1$.

- The medians of a triangle are concurrent (all pass through the same point); the point of concurrency trisects each median.

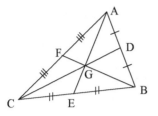

$$EG:EA = 1:3$$
$$DG:DC = 1:3$$
$$FG:FB = 1:3$$

- The altitudes of a triangle are concurrent.

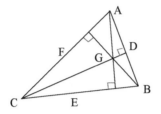

- The angle bisectors of a triangle are concurrent.
- The perpendicular bisectors of a triangle are concurrent; the point of concurrency is the centre of the circle which passes through the vertices of the triangle.

Related problems

You should be able to use this knowledge to solve the following kinds of problems.

- Finding the equation of a straight line given two points on the line or one point and the gradient.
- Find the perpendicular bisector of a line segment.

Remember that if a line passes through (a, b) and has a gradient of m its equation is $y - b = m(x - a)$

Given a line segment AB with endpoints $A(x_a, y_a)$ and $B(x_b, y_b)$

its midpoint C is $\left(\dfrac{x_a + x_b}{2}, \dfrac{y_a + y_b}{2}\right)$ and its gradient $m_{AB} = \dfrac{y_a - y_b}{x_a - x_b}$

C lies on the perpendicular bisector and has a gradient $m_\perp = -\dfrac{1}{m_{AB}}$

● Find medians, altitudes and perpendicular bisectors of triangles.

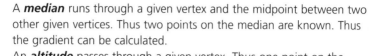

> A **median** runs through a given vertex and the midpoint between two other given vertices. Thus two points on the median are known. Thus the gradient can be calculated.
>
> An **altitude** passes through a given vertex. Thus one point on the altitude is known. It is at right angles to the opposite side whose gradient can be calculated. Hence its gradient is known.
>
> A **perpendicular bisector** passes through the midpoint of a side (so a point is known) at right angles (so its gradient can be calculated).

● Finding the intersection of two lines.
● Finding the angle at which two lines intersect.

> the line $y = ax + b$ intersects the line $y = Ax + B$ where $ax + b = Ax + B$

● finding the angle at which two lines intersect.

> A line with a gradient m will make an angle of $\tan^{-1}m$ with the positive direction of the x-axis.

> If the equations of two intersecting lines are known, their gradients can easily be obtained. Thus the angle each makes with the x-axis can be obtained. The two lines and the x-axis form a triangle. The angle of intersection of the two lines will be the third angle in the triangle. Note from the diagram that $a = \tan^{-1}m_2 - \tan^{-1}m_1$

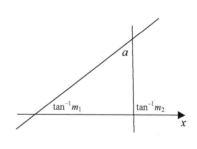

Objective questions

❶ Calculate the gradient of the line *perpendicular* to the line with equation $3x + 4y = 6$.

A $\dfrac{3}{4}$ B $-\dfrac{3}{4}$ C $\dfrac{4}{3}$ D $-\dfrac{4}{3}$

Rough working

In this case the strategy is to rearrange the equation into the form $y = mx + c$ and thereafter read off the gradient.

Your rough working might look like this:

$3x + 4y = 6$

$\Rightarrow 4y = -3x + 6$

$\Rightarrow y = -\dfrac{3}{4}x + \dfrac{6}{4}$

$m = -\dfrac{3}{4}$

The gradient of the line perpendicular to this one is $\dfrac{4}{3}$

Choose option C

❷ What is the equation of the line which passes through (1, 2) and (3, 4)?

    A $y = 2x$ B $y = 3x - 1$ C $y = x + 1$ D $y = x - 1$

Rough Working

In this case we could work out the gradient and then use
$y - y_1 = m(x - x_1)$ before finally rearranging it into the form $y = mx + c$.

However an easier way, given we have a list of possible answers, would be to substitute $x = 1$ and see which of the options give $y = 2$. Secondly substitute $x = 3$ and see which of the options give $y = 4$.

The option you want will fit both tests.

Only option C fits the description.

Choose option C

❸ Which of the following could be the graph of $x - y = 4$?

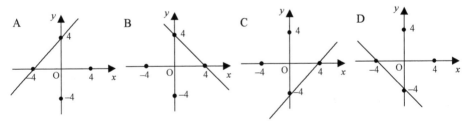

Rough Working

Rearranging the equation into the form $y = x - 4$ lets us see that the gradient is positive (so it must be A or C) and it cuts the y-axis at -4. So it can only be C.

Choose option C.

The answer to each of the preceding three questions is C. This is coincidence. Don't look for a pattern in the answers of the objective part of the test. There won't be a deliberate pattern.

An extended response question that doesn't need a calculator

❹ Triangle ABC has vertices A($-5,1$), B($-3, 5$) and C($3, -3$).
BM is the median of the triangle which passes through B.
M lies on AC.

a) Find the equation of BM.   *3 marks*
b) Find the equation of the perpendicular bisector of AB.   *4 marks*
c) Where does BM cut this perpendicular bisector?   *4 marks*

(**Response**)──

a) A median of a triangle is a line drawn from a vertex to the mid-point of the opposite side. M must be the midpoint of AC.

So coordinates of M are $\left(\dfrac{x_a + x_c}{2}, \dfrac{y_a + y_c}{2}\right) = \left(\dfrac{-5 + 3}{2}, \dfrac{1 + (-3)}{2}\right) = (-1, -1)$

The gradient of MB: $m_{MB} = \dfrac{y_b - y_m}{x_b - x_m} = \dfrac{5 - (-1)}{-3 - (-1)} = \dfrac{6}{-2} = -3$

The equation of BM: $y - (-1) = -3(x - (-1))$

We will be using this later in the question so it is worthwhile simplifying it.

$y + 1 = -3x - 3$ or $y = -3x - 4$ or $y + 3x = -4$

b) Coordinates of midpoint of AB are $\left(\dfrac{-5 + (-3)}{2}, \dfrac{1 + 5}{2}\right) = (-4, 3)$

$m_{AB} = \dfrac{5 - 1}{-3 - (-5)} = \dfrac{4}{2} = 2$

$m_\perp = -\dfrac{1}{2}$

Equation of perpendicular bisector of AB: $y - 3 = -\dfrac{1}{2}(x - (-4))$

Again we will use this equation and so simplification is useful ...

The equation simplifies to $2y + x = 2$

c) To find where BM cuts the perpendicular bisector we solve their equations simultaneously.

$$y + 3x = -4 \quad \dots \quad ①$$
$$2y + x = 2 \quad \dots \quad ②$$

① x 2 :
$$2y + 6x = -8 \quad \dots \quad ③$$
③ – ② :
$$5x = -10$$
$$\Rightarrow x = -2$$
Substitute in ① : $y - 6 = -4$
$$\Rightarrow y = 2$$

Thus the point of intersection is ($-2, 2$) ──────────────◯

How might the above responses attract marks?

In the following marking scheme, each bullet point attracts a mark; the superscripted number is just a reference number ... we can then refer to mark 1, mark 2 etc.

There may be more than one way to answer a question so the *ic*, *ss*, *pd* labels help in finding where the marks can be applied in other alternative solutions. Eventually the description of what to award the mark for will become less specific to help in making the scheme transferable.

Marking scheme

Part a)

- •[1] know to find mid-point   *ss*
- •[2] find gradient   *pd*
- •[3] state equation of line   *ic*

Although stating the equation might attract all three marks, you should emphasize your working when finding the mid-point and the gradient. Part-marks are then still available when an error is made. Although errors might occur in mark 1 and/or mark 2, mark 3 can be obtained when the working for it is correctly followed through and consistent with the errors.

Part b)

- •[4] know to find mid-point   *ss*
- •[5] find gradient of AB   *pd*
- •[6] find the gradient of the perpendicular   *ss*
- •[7] state equation of line   *ic*

The evidence for mark 6 *should be stated*. No one has given you the fact that $m_{line} \times m_{\perp} = -1$ and so a strategy mark will be awarded for this or an equivalent declaration.

Part c)

- •[8]   strategy for solving equations   *ss*
- •[9]   scaling equation in preparation for elimination   *pd*
- •[10] solve for *x*   *pd*
- •[11] solve for *y* and state the coordinates of the point   *ic*

The final mark is a communication mark and in general will be awarded when the information asked for is given in the right form.
Do not be tempted to plot the points accurately.
The subsequent drawing would count as a scale drawing and no credit can be given for answers obtained from it, for example the coordinates of the mid-points.

Extended response questions that need a calculator

5 Calculate the acute angle at which the lines $y = 2x - 3$ and $y = 2 - 3x$ intersect. *5 marks*

(Response)

When answering a question like this a sketch often helps.
Labelling will help in the efficient communication of your strategies.
The strategy itself depends on the knowledge that the gradient of a line making an angle of $\theta°$ with the positive direction of the x-axis (scales being equal) is $m = \tan\theta°$

In the diagram AC has the equation $y = 2 - 3x$.

$m_{AC} = -3$ by inspection.

$\tan(\angle ACx) = -3$

$\Rightarrow \angle ACx = \tan^{-1}(-3) = 108\cdot4°$

[The calculator gives $-71\cdot6$ for $\tan^{-1}(-3)$
... from the sketch we know we want $-71\cdot6 + 180$.]

So $\angle ACB = 71\cdot6°$.

Similarly AB has the equation $y = 2x - 3$.

$m_{AB} = 2$ by inspection.

$\tan(\angle ABC) = 2$

$\Rightarrow \angle ABC = \tan^{-1}(2) = 63\cdot4°$

Thus $\angle BAC = 180° - 71\cdot6° - 63\cdot4° = 45°$

The lines intersect at 45°

Marking scheme

•¹	interpret equation to get gradient	*ic*
•²	know the relation between gradient and angle	*ss*
•³	solve trig equation for angle	*pd*
•⁴	repeat steps for second angle	*pd*
•⁵	find third angle of a triangle	*ss*

21

6 Calculate the area of the triangle with vertices A(–1, –2), B(5, 1) and C(1, 4).   *6 marks*

(*Response*)

More often than not such a question will be broken down, providing sub-questions that steer you through a suitable strategy. However, if left in the form above, the selection of strategy will carry weight in the marking scheme. You should try to be very clear in your route through to a solution.

There are a couple of possible main strategies.

Strategy 1

- Find the length of a side, say AB.
- Find the equation of AB.
- Find the equation of the altitude from C.
- Find where the altitude cuts AB.
- Find the length of the altitude.
- Area = ½ × AB × altitude.

Strategy 2.

- Find the length of the three sides.
- Use the cosine rule to find an angle.
- Use Area = ½*ab* sin C.

Strategy 2 offers the shortest road to the answer.

State the distance formula at least once:

$$AB = \sqrt{(x_B - x_A)^2 + (y_B - y_A)^2}$$

$$AB = \sqrt{(5 - (-1))^2 + (1 - (-2))^2} = 6{\cdot}708$$

$$AC = \sqrt{(1 - (-1))^2 + (4 - (-2))^2} = 6{\cdot}325$$

$$BC = \sqrt{(1 - 5)^2 + (4 - 1)^2} = 5$$

State the cosine rule
… and a sketch would help.

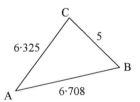

$$\cos B = \frac{5^2 + 6{\cdot}708^2 - 6{\cdot}325^2}{2 \times 5 \times 6{\cdot}708} = 0{\cdot}477$$

$$\angle B = 63{\cdot}4°$$

Now use the area rule:

Area = 0·5 × 5 × 6·708 × sin 63·4° = 15·0 (to 3 s.f.)

Marking scheme

- $\bullet^1$ interpret the situation   *ic*
- $\bullet^2$ use the distance formula to find the sides   *pd*
- $\bullet^3$ know to use the cos rule: *ss*
- $\bullet^4$ find the angle B   *pd*
- $\bullet^5$ know to use sine-area rule   *ss*
- $\bullet^6$ find the area   *pd*

Here is an example of where a mixing of topics and strategies occurs.
This particular topic, the straight line, is likely to appear in questions involving gradients, viz the circle, polynomials and differentiation.

> Often a question will ask you to provide the **exact** value for an answer. Often the word is emboldened and in italics to draw this to your attention. When this is the case marks will be lost whenever you evaluate a surd or trigonometric ratio using your calculator. You must display your ability to handle surds and trig identities.

Take, for example, the previous question. It could have been written as:

Calculate, *exactly*, the area of the triangle with vertices A(–1, 2), B(5, 1) and C(1, 4).

State the distance formula at least once:

$$AB = \sqrt{(x_B - x_A)^2 + (y_B - y_A)^2}$$
$$AB = \sqrt{(5 - (-1))^2 + (1 - (-2))^2} = \sqrt{45} = 3\sqrt{5}$$
$$AC = \sqrt{(1 - (-1))^2 + (4 - (-2))^2} = \sqrt{40} = 2\sqrt{10}$$
$$BC = \sqrt{(1 - 5)^2 + (4 - 1)^2} = 5$$

State the cosine rule
... and a sketch would help.

$$\cos B = \frac{5^2 + \sqrt{45}^2 - \sqrt{40}^2}{2 \times 5 \times 3\sqrt{5}} = \frac{1}{\sqrt{5}}$$

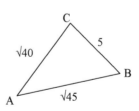

$\angle B = ...$ We don't actually need the size of the angle. We just need its sine.

Since it is an angle in a triangle we can use

$$\sin B = \sqrt{1 - \cos^2 B} = \sqrt{1 - \frac{1}{5}} = \sqrt{\frac{4}{5}} = \frac{2}{\sqrt{5}}$$

> Remember from Standard Grade that $\sin^2 A + \cos^2 A = 1$

Now use the area rule:

Area $= 0{\cdot}5 \times 5 \times 3\sqrt{5} \times \sin B° = 0{\cdot}5 \times 5 \times 3\sqrt{5} \times \dfrac{2}{\sqrt{5}} = 15$ (exactly).

The marking scheme would be the same but with a note that marks would be lost when a decimal approximation of a surd or a trig ratio is used.

In this form, the question offers 6 marks for the person who can see the complete strategy. 6 marks are too many lose if this strategy is not apparent. As said before, for this reason it is more likely that the question would be expanded to allow a C-skills entry into the question.

For example,

A triangle has vertices A(−1, 2), B(5, 1) and C(1, 4).

(a)   By considering AB^2, AC^2 and BC^2 decide if the triangle is right-angled.

(b)   Find the *exact* value of cos B.

(c)   Find the *exact* value of the area of the triangle.

Topic Tips

As already said, in mathematics your grade depends on how many marks you pick up and there are only three places to pick up marks ... ic, ss, pd.

In this topic marks are commonly lost by a lack of communication.

State clearly you are finding a gradient; state clearly that if AB and AC are perpendicular then $m_{AB}.m_{AC} = -1$; state clearly the equation of any line in the form $y - y_1 = m(x - x_1)$ before attempting any simplification.

5 Functions and Graphs

What you should know

Related problems

Objective questions

Extended response questions that don't need a calculator

What you should know

You are expected to know the following facts.

- A **function**, f, is a rule which maps every member of one set (the **domain**) onto another set (the **range**).

  Each member of the domain, x, maps onto its **image** $f(x)$ in the range.

  Each member of the domain has only one image.

  The set of images is called the **range** of the function.

  A rule which takes each member of the range of f and maps it back to the original domain is called the **inverse** of the function and is denoted by f^{-1}.

  $f^{-1}(f(x)) = x$

  When a function, g, maps the members of the range of f onto another set, a composite function is formed. Each x of the domain of f has an image $g(f(x))$ in the range of g.

- The features of graphs and the probable form of a function from its graph.

  (i) $y = a_n x^n + a_{n-1} x^{n-1} + a_{n-2} x^{n-2} + \ldots + a_2 x^2 + a_1 x + a_0 ; \ a_n \neq 0$

  (a polynomial of degree n)

  Examples: $ax^2 + 3x + 4$ is a polynomial of degree 2 as long as $a \neq 0$.

  $7x^3 + 2x^2 + 5$ is a polynomial of degree 3 … notice $7 \neq 0$.

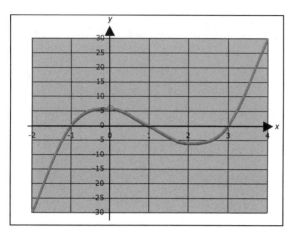

If you are told this is a graph of a cubic function passing through (−1, 0), (0, 6),(1, 0) and (3, 0), then since it cuts the x-axis in three places, it will have three factors of the form $(x - b)$ where b is an x-intercept.

Its equation will be of the form $y = a(x - b)(x - c)(x - d)$.

Since it cuts at $x = -1$, $x = 1$ and $x = 3$, the form becomes $y = a(x + 1)(x - 1)(x - 3)$.

Since $x = 0$ when $y = 6$ we get $6 = a(0 + 1)(0 - 1)(0 - 3) \Rightarrow 3a = 6 \Rightarrow a = 2$.

Thus the graph has the equation $y = 2(x + 1)(x - 1)(x - 3)$.

You will learn more of this in Unit 2.

(ii) $y = \sin(ax + b)$,

> In Higher mathematics a new unit of measure for angles, radians, was introduced.
> π radians = 180°

Don't mix your units. Marks will be lost if you work in degrees when the question uses radians. If no sign suggests otherwise then you must work in radians: $y = \sin(ax + b)$ ... no sign ... so radians but $y = \sin(ax + b)°$ ... degrees.

Example:

> In any sine wave, with the equation in the form $y = A\sin(Bx + C) + D$
> A ... the *amplitude* ... equal to half the difference between the
> maximum and minimum values.
> B ... the *frequency* ... equal to the number of waves in 2π radians.
> C ... the *phase* ... equal to the number of radians the wave has been
> *shifted* to the left given that the basic sine wave passes through (0, 0).
> D ... the *y-displacement* ... equal to the amount of shift in the
> y-direction, given that in the sine wave the x-axis lies midway between
> the maximum and minimum values.
> In appendix 5 there are instructions on how to make a spreadsheet to
> explore this function more dynamically. (See *www.leckieandleckie.co.uk*.)

$y = \sin(ax + b)$

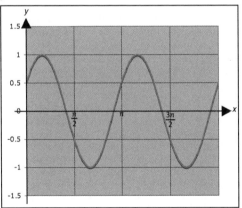

In this graph we manage **two** complete waves in 2π radians ... so $a = 2$.
The curve cuts the y-axis when $x = 0$ ... when $\sin(0 + b) = 0{\cdot}5$ ... when
$b = \sin^{-1}(0{\cdot}5) = \dfrac{\pi}{6}$
The curve has equation $y = \sin(2x + \dfrac{\pi}{6})$.

(iii) $y = \cos(ax + b)$

In this graph we manage three complete waves in 2π radians ... so $a = 3$.
The curve cuts the y-axis when $x = 0$ ... when $\cos(0 + b) = 0{\cdot}5$ ... when
$b = \cos^{-1}(0{\cdot}5) = \dfrac{\pi}{3}$

The curve has equation $y = \cos(3x + \dfrac{\pi}{3})$.

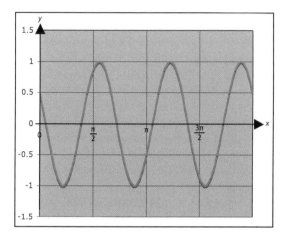

(iv) $y = a^x$ (both $a > 1$ and $0 < a < 1$, $x \in R$)

> Life amongst the logs and exponentials is easier if you get into the habit of mentally switching between forms, e.g. $y = a^x$ is equivalent to $\log_a y = x$.
> Both are describing the same relationship between x and y.

Example ($a > 1$)

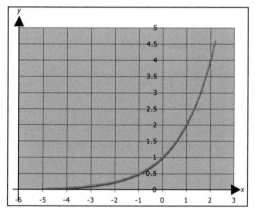

Note that when $x = 0$ then $y = a^0 = 1$.
All curves of this kind will pass through (0, 1).
In this case when $x = 1$ then $y = 2$ ... $2 = a^1$
... $a = 2$.
The curve has equation $y = 2^x$.

Example (0 < a < 1)

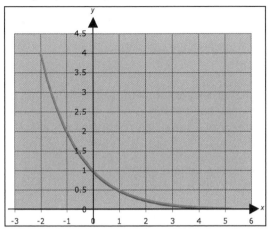

Again note that when $x = 0$ then $y = a^0 = 1$.
All curves of this kind will pass through (0, 1).
In this case when $x = 1$ then $y = 0.5$ ... $0.5 = a^1$
... $a = 0.5$.
The curve has equation $y = 0.5^x$.

(v) $y = \log_a x$ $(a > 1, x > 0)$

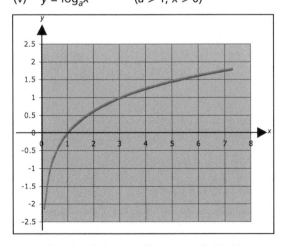

Note that (1, 0) lies on all curves of this kind ... $\log_a 1 = 0$.
In this case, the curve passes through (3, 1) ... $\log_a 3 = 1 \Leftrightarrow a = 3$.
The curve has equation $y = \log_3 x$.

- $x^2 + 2ax = (x + a)^2 - a^2$ ... a grade C skill ... completing the square with the coefficient of the x^2 term equal to 1.

> This identity comes directly from the simple expansion
> $(x + a)^2 = x^2 + 2ax + a^2$.

Its usefulness lies in the fact that an expression with two terms in x has been reduced to an expression with only one term in x.
If the coefficient of the x^2 term is not 1 the skill is considered a grade A/B skill (see later).

- The definition of radian measure ... π radians = 180°.
If no units are mentioned or if calculus is the context you must assume that radian measure is being used.

Why?

When working in calculus we have to consider the ratio $\frac{\sin x}{x}$ for small values of x.

When working in degrees and with small values of x we find $\frac{\sin x°}{x°} \approx \frac{\pi}{180}$.

When working in radians and with small values of x we find $\frac{\sin x}{x} \approx 1$.

This simplifies things greatly when working with calculus.

Note that using degrees is a Standard Grade skill.

- The exact values of the sine, cosine and tangent of angles associated with:

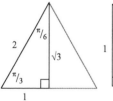
Half-equilateral triangle

Half-square

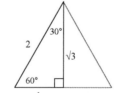
Half-equilateral triangle

Half-square

These crop up with great regularity in the exams.

> When given a question in radians, give your answer in radians or risk losing marks.

Related problems

You should be able to use the above information to solve the following kinds of problems.

- Drawing the graphs of related functions given the graph of *f(x)*.

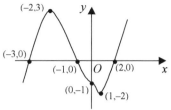

Example y = f(x)

C grade:

(i) $y = -f(x)$ ... reflection in the *x*-axis.
(ii) $y = f(-x)$ ... reflection in the *y*-axis.

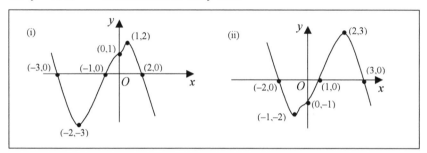

(iii) $y = f(x) + a$ ... a translation of *a* units in the *y*-direction.
                      Add *a* to all your *y*-coordinates.
(iv) $y = f(x + a)$ ... a translation of *–a* units in the *x*-direction.
                      Subtract *a* from all your *x*-coordinates.

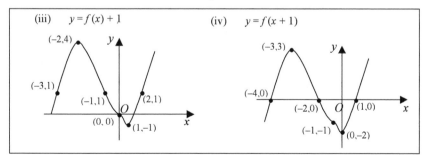

(v) $y = af(x)$ ... a stretching of the curve by a factor of *a* in the *y*-direction.
                    Multiply all your *y*-coordinates by *a*.
(vi) $y = f(ax)$ ... a squashing of the curve by a factor of *a* in the *x*-direction.
                    Divide all your *x*-coordinates by *a*.

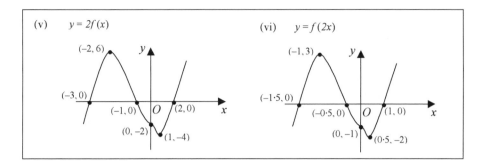

(v) $y = 2f(x)$

(vi) $y = f(2x)$

A/B grade:

(vii) $y = af(x) + b$
(viii) $y = f(ax + b)$
(ix) $y = -af(x)$
(x) $y = af(x + b)$

> If the given graph contains specific points whose coordinates are given then your related graph should, where possible, have the coordinates of the corresponding point.
> This is often where the evidence for the mark will be found.

- Finding composite functions of the form $f(g(x))$, given $f(x)$ and $g(x)$.

  Example

  $f(x) = 2x + 1$ and $g(x) = x^2 + 2x - 3$. Find expressions for $f(g(x))$ and $g(f(x))$

  (Response) ────────────────────────────────

  $f(g(x)) = f(x^2 + 2x - 3) = 2(x^2 + 2x - 3) + 1 = 2x^2 + 4x - 5$

  $g(f(x)) = g(2x + 1) = (2x + 1)^2 + 2(2x + 1) - 3 = 4x^2 + 4x + 1 + 4x + 2 - 3 = 4x^2 + 8x$

  ────────────────────────────

  In general $f(g(x)) \neq g(f(x))$ though you may be asked to find the value(s) of x for which $f(g(x))$ does equal $g(f(x))$ ... or to prove no such x exists.

  For example

  $f(x) = 2x + 1$ and $g(x) = x^2 + 2x - 3$

  Show that for no values of x does $f(g(x)) = g(f(x))$.

(**Response**)

$f(g(x)) = 2x^2 + 4x - 5$ ... see above

$g(f(x)) = 4x^2 + 8x$

$f(g(x)) = g(f(x))$

$\Rightarrow 2x^2 + 4x - 5 = 4x^2 + 8x$

$\Rightarrow 2x^2 + 4x + 5 = 0$

The discriminant of this equation is $4^2 - 4.2.5 = 16 - 40 < 0$.
So no real roots exist for this equation.
So there is no value of x for which $f(g(x)) = g(f(x))$. ——————

- Sketching and annotating the graphs of functions once the critical features (i.e. x-intercepts, y-intercept, axes of symmetry, stationary points, behaviour at infinity or end-points) have been identified.
  [See chapter 6 for an example.]

- Completing the square in a quadratic of the form $ax^2 + bx + c = 0$.
  Note that $a = 1$ for C-grade skills and when $a \neq 1$ it is considered a grade A/B skill.

  Example (grade C)
  $x^2 + 8x + 2 = (x + 4)^2 - 4^2 + 2 = (x + 4)^2 - 14$

  Example (grade A/B)
  $3x^2 + 8x + 2 = 3(x^2 + \frac{8}{3}x) + 2 = 3((x + \frac{4}{3})^2 - (\frac{4}{3})^2) + 2 = 3(x + \frac{4}{3})^2 - \frac{10}{3}$

  Note: Many schools leave the coverage of this topic and the following material until unit 2.

- Deduce the minimum value of a quadratic by first completing the square and the value of x where it occurs.

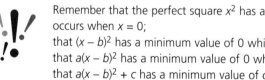
Remember that the perfect square x^2 has a minimum value of 0 which occurs when $x = 0$;
that $(x - b)^2$ has a minimum value of 0 which occurs when $x = b$;
that $a(x - b)^2$ has a minimum value of 0 which occurs when $x = b$;
that $a(x - b)^2 + c$ has a minimum value of c which occurs when $x = b$.

Again, if completing the square involves a problem with a coefficient different from 1 then it will be considered grade A/B.

Example (grade C)
Find the minimum value of $x^2 + 4x + 3$ and the corresponding value of x .

(Response) ─────────────────────────────────────

$x^2 + 4x + 3 = (x + 2)^2 - 4 + 3 = (x + 2)^2 - 1$
The minimum value of $(x + 2)^2$ is 0 when $x = -2$.
So the minimum value of $(x + 2)^2 - 1$ is -1 when $x = -2$. ───────⬤

Example (grade A/B)
Find the maximum value of $3 - 4x - x^2$ and the value of x at which it occurs.

(Response) ─────────────────────────────────────

$3 - 4x - x^2 = 3 - (x^2 + 4x) = 3 - [(x + 2)^2 - 4] = 3 - (x + 2)^2 + 4 = 7 - (x + 2)^2$
The minimum value of $(x + 2)^2$ is 0 when $x = -2$.
So the maximum value of $7 - (x + 2)^2$ is 7 when $x = -2$. ───────⬤

- Interpret formulae and equations identifying maxima and minima.
  This can be assisted by knowing $x^2 \geq 0$; $-1 \leq \sin x \leq 1$; $-1 \leq \cos x \leq 1$.

> (i) $a - x$ will be at its maximum when x is at its minimum.
> (ii) $\frac{a}{x}$ will be at its maximum when x is at its minimum.
> (iii) $-a + b \leq a\sin(f(x)) + b \leq a + b$

Example
Find the maximum value of $\dfrac{3}{x^2 + 2x + 6}$ and the value of x at which this maximum occurs.

(Response) ─────────────────────────────────────

From note (ii) we can see that the maximum will occur when the quadratic expression is at a minimum.
$x^2 + 2x + 6 = (x + 1)^2 - 1 + 6 = (x + 1)^2 + 5$
When $x = -1$ this will have a minimum value of 5.
Thus the original expression will have a maximum of $\frac{3}{5}$ when $x = -1$. ───⬤

Trigonometric functions are also likely subjects for this type of question.

Example
What is the minimum value of $3\sin(2x - \frac{\pi}{2}) + 5$
and for what values of x, $0 \leq x \leq 2\pi$, does it occur?

The minimum value of the sine function is -1. This occurs when the angle is $\frac{3\pi}{2}$.
So the minimum value of the given function is $3 \times (-1) + 5 = 2$.
This occurs when $2x - \frac{\pi}{2} = \ldots -\frac{\pi}{2}, \frac{3\pi}{2}, \frac{7\pi}{2}, \frac{11\pi}{2}$
         when $2x = \ldots 0, \frac{4\pi}{2}, \frac{8\pi}{2}, \frac{12\pi}{2}$
         when $x = \ldots 0, \frac{4\pi}{4}, \frac{8\pi}{4}, \frac{12\pi}{4}$.
In the given range, when $x = 0, \pi, 2\pi$.

<u>Note</u>

(i)   When first listing our possibilities we had to list $\frac{3\pi}{2} \pm 2n\pi$ (i.e. $\frac{3\pi}{2} \pm$ *any number of revolutions*), including one negative value ... we should think of this when a later step will involve the *addition* of an angle to each member of our list ... possibly 'pulling' it into our range as happened here.

(ii)   We anticipate dividing by 2 since we have $2x$. So we include in our initial list values that fall in the range 0 to 4π.

If the expression had included $3x$ then we would list from 0 to 6π etc.

Objective questions

1   Here is a sketch of the graph $y = f(x)$.

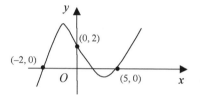

Here is a sketch of the graph *of a related function.*

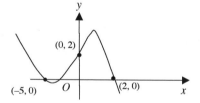

Of which of the following is it most likely to be a sketch?

A $y = f(-x)$ B $y = -f(x)$ C $y = f(x) - 3$ D $y = f(x) + 3$

Rough working

An inspection of the curve and the points it passes through show that reflection in the y-axis has taken place. So $y = f(-x)$.

Choose option A

❷   Here is a graph that has an equation of the form $y = \log_a bx$.
What are the values of a and b?

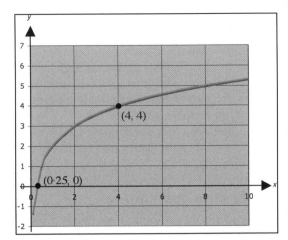

A $a = 0\cdot5$, $b = 2$
B $a = 2$, $b = 0\cdot5$
C $a = 2$, $b = 4$
D $a = 4$, $b = 2$

Rough working

$(0\cdot25, 0)$ lies on graph $\Rightarrow \log_a \frac{b}{4} = 0 \Rightarrow \frac{b}{4} = 1 \Rightarrow b = 4$
$(4, 4)$ lies on graph $\Rightarrow \log_a b \times 4 = 4 \Rightarrow \log_a 16 = 4$ (since $b = 4$)
$\Rightarrow 16 = a^4 \Rightarrow a = 2$

$\log_a x = y \Leftrightarrow x = a^y$

Choose option C

If possible, look for some confirmation of your choice.
The curve appears to pass through $(2, 3)$
     ... $\log_2(4 \times 2) = \log_2 8 = 3$ ... confirmed.

❸ $f(x) = 3x - 2$ and $g(x) = x^2 - 1$
Find an expression for $g(f(x))$.

A $9x^2 - 12x + 3$
B $9x^2 - 12x - 5$
C $3x^2 - 5$
D $3x^2 + 3$

Rough working

$g(f(x)) = g(3x - 2) = (3x - 2)^2 - 1 = 9x^2 - 12x + 4 - 1 = 9x^2 - 12x + 3$

Choose option A

For a question such as this, a useful check on your answer would be to quickly examine special cases e.g. $x = 0$.
$f(0) = 3 \times 0 - 2 = -2$; $g(-2) = (-2)^2 - 1 = 3$ i.e. $g(f(0)) = 3$
Now put $x = 0$ into your four options.

... A gives 3; B gives –5; C gives –5 and D gives 3
Now try $x = 1$
$f(1) = 3 \times 1 - 2 = 1$; $g(1) = (1)^2 - 1 = 0$ i.e. $g(f(1)) = 0$
Now put $x = 1$ into your four options.
... A gives 0; B gives –8; C gives –2 and D gives 6,
confirming A as the correct choice.

4 Express $4x^2 + 8x - 1$ in the form $a(x + b)^2 + c$

A $4(x + 4)^2 - 65$
B $4(x + 1)^2 - 2$
C $4(x + 2)^2 - 17$
D $4(x + 1)^2 - 5$

Rough working

$4x^2 + 8x - 1 = 4(x^2 + 2x) - 1 = 4[(x + 1)^2 - 1^2] - 1 = 4(x + 1)^2 - 4 - 1 = 4(x + 1)^2 - 5$

Choose option D

This is a level A/B type question. It is important to check your choice.
Again, examining a special case helps.

Example

$x = 0$ ... The expression gives $4x^2 + 8x - 1 = 0 + 0 - 1 = -1$
... **A gives –1**; B gives 2; **C gives –1** and **D gives –1**
Now try $x = -1$ ... The expression gives $4x^2 + 8x - 1 = 4 - 8 - 1 = -5$
... A gives –29; B gives –2; C gives –13 and **D gives –5**,
confirming D as the correct choice.

> Roughly 25% of the objective questions will be aimed at A or B grade.

5 $f(x) = 4 - 3\sin(2x - \frac{\pi}{4})$, $0 \le x \le \pi$.

What is the maximum value of $f(x)$ and for what value of x does it occur?

A $f_{max} = 7$ when $x = \frac{3\pi}{8}$
B $f_{max} = 7$ when $x = \frac{7\pi}{8}$
C $f_{max} = 1$ when $x = \frac{3\pi}{8}$
D $f_{max} = 1$ when $x = \frac{7\pi}{8}$

Rough working

$f(x)$ is the difference of two terms. It will achieve its maximum value
when the second term is at its minimum.
$3\sin(2x - \frac{\pi}{4})$ is at a minimum when $\sin(2x - \frac{\pi}{4})$ is at a minimum.

The sine of an angle will first achieve its minimum, −1, when the angle is $\frac{3\pi}{2}$.

$f(x)$ has a maximum value of $4 − (−3) = 7$ when $2x − \frac{\pi}{4} = \frac{3\pi}{2}$.

i.e. when $x = \frac{7\pi}{8}$

Choose option B

Check by putting each of the x-values into the function.

A $f(\frac{3\pi}{8}) = 1$ (not 7);   B $f(\frac{7\pi}{8}) = 7$;   C $f(\frac{3\pi}{8}) = 1$;   D $f(\frac{7\pi}{8}) = 7$ (not 1)

Extended response questions that don't need a calculator

⑥ The profit made per person by a bus company is dependent on the average speed, x mph, the drivers are instructed to maintain.

The profit, £R per person, is calculated using the formula:

$$R(x) = \frac{100}{x^2 − 100x + 3000}$$

(a) Express $x^2 − 100x + 3000$ in the form $a(x + b)^2 + c$.

(b) Hence calculate the maximum profit and the recommended average speed at which this can be attained.

(Response)

(a) $x^2 − 100x + 3000 = (x − 50)^2 − 50^2 + 3000 = (x − 50)^2 + 500$

(b) $x^2 − 100x + 3000$ reaches a minimum of 500 when $x = 50$

R reaches a maximum at the same time.

$R_{max} = \frac{100}{500} = 0 \cdot 2$ when $x = 50$

Maximum profit per person = £0·20 when the bus travels at an average speed of 50 mph.

Marking scheme

●¹ completing the square   *pd*
●² identify crucial aspect: R at max when (a) at min   *ss*
●³ make deductions: R_{max} and corresponding x *ss*
●⁴ express solution in language appropriate to situation   *ic*

Note that the question is framed in a context. The solution should be reported within the same context. In some circumstances this could affect the accuracy to which the answer is given or the units.

7 The diagram shows a sketch of $y = a \times 2^x$; $0 \le x \le 5$.
The point $(3, 1 \cdot 6)$ lies on this curve.

(a) What is the value of a?

(b) Under a particular scheme, a loan of £6 400 is paid back over five years. The amount of money owing, £y thousands, after x years is given by the function $y = 6 \cdot 4 - a \times 2^x$ where a takes the same value as in part (a).

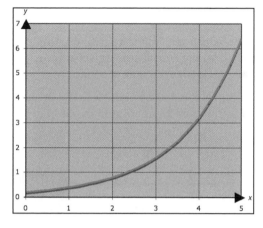

   (i) Make a sketch of this function.
   (ii) After how many years is half the loan paid back?

(**Response**)

(a) Using the point $(3, 1 \cdot 6)$ we get $1 \cdot 6 = a \times 2^3 \Rightarrow 1 \cdot 6 = 8a \Rightarrow 0 \cdot 2 = a$

(b) (i) Take two steps:
   step 1: $y = -0 \cdot 2 \times 2^x$ ... a reflection in the x-axis.
   step 2: $y = -0 \cdot 2 \times 2^x + 6 \cdot 4$ ... a translation in the y-direction of $6 \cdot 4$ units.

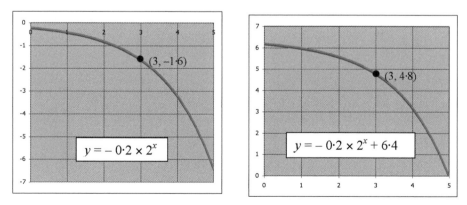

   (ii) $-0 \cdot 2 \times 2^x + 6 \cdot 4 = 3 \cdot 2 \Rightarrow 3 \cdot 2 = 0 \cdot 2 \times 2^x \Rightarrow 2^x = 16 \Rightarrow x = 4$
   After four years half the loan is repaid.

Marking scheme

- •¹   identify crucial aspects: substitute   *ss*
- •²   solve for *a*   *pd*
- •³   use symmetry   *ss*
- •⁴   sketch and annotate final graph   *ic*
- •⁵   identify crucial aspects: substitute   *ss*
- •⁶   solve an exponential equation   *ic*

Sketches should be annotated. The image of any annotated point in the original sketch should be annotated in the final sketch.

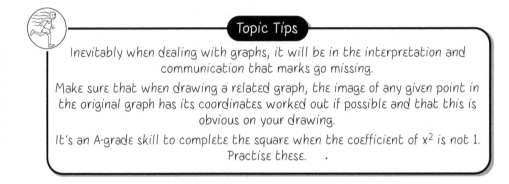

Topic Tips

Inevitably when dealing with graphs, it will be in the interpretation and communication that marks go missing.

Make sure that when drawing a related graph, the image of any given point in the original graph has its coordinates worked out if possible and that this is obvious on your drawing.

It's an A-grade skill to complete the square when the coefficient of x^2 is not 1. Practise these.   .

6 | Differential Calculus

What you should know

You are expected to know the following facts.

Basic differentiation

- A function of x has a **limit**, L, if as x tends towards some value a, the function tends towards a value of L. Note $f(a)$ need not exist for this limit to exist.

Example

Using smaller and smaller values of x, and with the calculator set to radians, explore the value of $\frac{\sin x}{x}$. You should find that the value of the function gets closer and closer to 1 as you let x approach zero. However, putting $x = 0$ will generate an error report.

We write $\lim\limits_{h \to 0} \frac{\sin x}{x} = 1$.

> This limiting value of 1 is dependant on working in radians.
> The limit is not 1 if working in degrees.

- A function $f(x)$ is **differentiable** at $x = a$ if the limit $\lim\limits_{h \to 0} \dfrac{f(a + h) - f(a)}{h}$ exists.

  This broadly means that if you can get the gradient of the tangent to the curve at $x = a$ then we say it is differentiable at $x = a$.

  It is often useful to restrict the domain of a function to some particular interval.

  If the function is differentiable at all points in the restricted domain we would use the expression 'differentiable over the interval'.

- To **differentiate** a function broadly means to derive or produce a function which will give us the gradient at any point on the graph of the function. This will also be the gradient of the tangent to the curve at that point.

- The **derivative**, or **derived function**, of a function $f(x)$ is defined as

  $f'(x) = \lim\limits_{h \to 0} \dfrac{f(x + h) - f(x)}{h}$.

  For functions of the form $f(x) = ax^n$ we find that this always turns out to be $f'(x) = anx^{n-1}$.

- The derivative tells us how y changes as x changes, the rate of change of y with respect to x.

  For this and other reasons the notation $\dfrac{dy}{dx}$ is also used as a symbol for $f'(x)$.

- *Derivatives follow certain rules.*

  (i)   If $f(x) = x^n$, then $f'(x) = nx^{n-1}$, $n \in Q$.
  (ii) If $f(x) = g(x) + h(x)$, then $f'(x) = g'(x) + h'(x)$.
  (iii) If $f(x) = kg(x)$, then $f'(x) = kg'(x)$ where k is a constant.

- How one variable changes with respect to another is referred to as the **rate of change** of the one with the other.

  The word 'per' will often be found associated with a rate of change in real-life contexts, e.g. km per litre, metres per second, tomatoes per kilo.

- Let $(x, f(x))$ and $(x + h, f(x + h))$ be two points on the curve $y = f(x)$.

  The average gradient between these points is defined as the gradient of the chord joining the points *viz* $\dfrac{f(x + h) - f(x)}{x + h - x} = \dfrac{f(x + h) - f(x)}{h}$.

- If $f'(x) > 0$ over some interval we say the function $f(x)$ is strictly increasing over the interval.

- If $f'(x) < 0$ over some interval we say the function $f(x)$ is strictly decreasing over the interval.

  The examiner can see if you can deal with this sort of analysis by asking questions like:

Show that the function $f(x) = x^3 - 6x^2 + 12x + 1$ is never decreasing.

(Response)

$f(x) = x^3 - 6x^2 + 12x + 1$
$f'(x) = 3x^2 - 12x + 12$
$\quad\ = 3(x - 2)^2$

A perfect square can never be negative and so $f'(x)$ can never be negative. Hence $f(x)$ is never decreasing.

How should you know to factorise?

The examiner will not ask trick questions. If you are asked to show that a function is never decreasing then it won't be. It stands to reason that the function can be expressed in a form that highlights this fact.

When asked questions about whether an expression is always positive or never negative etc. you should consider that:

(i) perfect squares are never negative;

(ii) $-1 \leq \sin x \leq 1$ and so a function like $2 + \sin x$ will always be positive ... lying between 1 and 3 as it does ... $1 \leq 2 + \sin x \leq 3$. ────────◯

- If $f'(x) = 0$ at some point we say the function $f(x)$ is stationary at that point. The value of $f(x)$ at this point is called the stationary value.

  (i) When the function is increasing before and decreasing after the stationary point we say the point is a maximum turning point (and the function will have a maximum turning value).

  (ii) When the function is decreasing before and increasing after the stationary point we say the point is a minimum turning point (and the function will have a minimum turning value).

  (iii) When the function is increasing before and increasing after the stationary point we say the point is a horizontal point of inflexion.

  (iv) When the function is decreasing before and decreasing after the stationary point we also say the point is a horizontal point of inflexion.

Example 1 $f(x) = x^2 + 2$
$\qquad\qquad f'(x) = 2x$
At $x = 0$, $f(0) = 2$ and $f'(0) = 0$ ...
(0, 2) is a stationary point.
For $x < 0$, $f'(x) < 0$ ... a decreasing function.
For $x > 0$, $f'(x) > 0$ ... an increasing function.
(0, 2) is a minimum stationary point.
The minimum stationary value of $f(x) = f(0) = 2$.

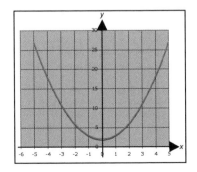

Example 2 $f(x) = x^3 + 6x^2 + 12x + 11$
                $f'(x) = 3x^2 + 12x + 12$
At $x = -2$, $f(-2) = 3$ and $f'(-2) = 0$ ...
$(-2, 3)$ is a stationary point.
For $x < -2$, $f'(x) > 0$ ... an increasing function.
For $x > -2$, $f'(x) > 0$ ... an increasing function.
$(-2, 3)$ is a horizontal point of inflexion.

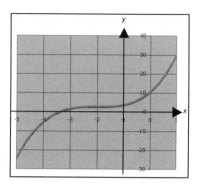

Related problems

You should be able to use this knowledge to solve the following kinds of problems.

- Finding the derived function of sums of terms where the terms are of the form $f(x) = ax^n$.
  You would be expected to simplify prior to differentiation.

  (a)   products
       Example 1: $f(x) = (2x + 1)^2 = 4x^2 + 4x + 1 \Rightarrow f'(x) = 8x + 4$
       Example 2: $f(x) = (3x - 5)(4 - x) = -20 + 17x - 3x^2 \Rightarrow f'(x) = 17 - 6x$

  (b)   rational expressions
       Example 1: $f(x) = \dfrac{x^2 + 2}{x} = \dfrac{x^2}{x} + \dfrac{2}{x} = x + 2x^{-1} \Rightarrow f'(x) = 1 - 2x^{-2}$
       Example 2: $f(x) = \dfrac{(x + 2)(2x - 3)}{3x} = \dfrac{2x^2 + x - 6}{3x} = \frac{2}{3}x + \frac{1}{3} - 2x^{-1}$
                  $\Rightarrow f'(x) = \frac{2}{3} + 2x^{-2}$

  (c)   surds
       Example 1: $f(x) = \sqrt{x} = x^{\frac{1}{2}} \Rightarrow f'(x) = \frac{1}{2}x^{-\frac{1}{2}}$
       Example 2: $f(x) = \dfrac{x + 1}{\sqrt{x}} = \dfrac{x + 1}{x^{\frac{1}{2}}} = x^{\frac{1}{2}} + x^{-\frac{1}{2}} \Rightarrow f'(x) = \frac{1}{2}x^{-\frac{1}{2}} - \frac{1}{2}x^{-\frac{3}{2}}$

  Any reasonable combination of the above could appear in the exam.

Remember the laws of indices. They are essential for this part of the work.
(i) $a^m \times a^n = a^{m+n}$
(ii) $\dfrac{a^m}{a^n} = a^m \div a^n = a^{m-n}$
     When $n = m$ this leads to $1 = a^0$.
     When $m = 0$ this leads to $\dfrac{1}{a^n} = a^{-n}$.
(iii) $(a^m)^n = a^{mn}$
     When $m = \frac{1}{n}$ this leads to $a^{\frac{1}{n}} = \sqrt[n]{a}$ and in particular $a^{\frac{1}{2}} = \sqrt{a}$.
     A combination of the above gives $a^{\frac{m}{n}} = \sqrt[n]{a^m} = (\sqrt[n]{a})^m$.

- Finding rates of change knowing that $f'(a)$ is the rate of change of f at a, e.g. in the context of motion where time is the independent variable, or expanding spheres where the radius is the independent variable.

Possible problems can be dressed up in context.

Given that the displacement of a particle from the origin, s cm, is a function of time, t seconds i.e. $s = f(t)$ then the rate of change of displacement with time, velocity, v cm/s is given by the derived function $v = \dfrac{ds}{dt} = f'(t)$

and the rate of change of velocity with time, acceleration, a cm/s² is given by the derived function $a = \dfrac{dv}{dt} = \dfrac{d^2s}{dt^2} = f''(t)$.

Example 1

A satellite is dropping back to Earth. The distance it has fallen, x metres, since first spotted is given by $x = 100t + 5t^2$, where t is the length of time in seconds since first spotted.

(a) How far did it travel in the first second?

(b) How fast was it travelling after ten seconds?

(c) What is the satellite's acceleration?

(d) The satellite will break up when its speed exceeds 500 m/s. How far will it have travelled before this happens?

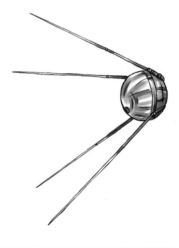

(Response)

(a) $t = 1 \Rightarrow x = 100 + 5 = 105$ metres

(b) $v = \dfrac{dx}{dt} = 100 + 10t$

   $t = 10 \Rightarrow v = 100 + 10 \times 10 = 200$ m/s

(c) $a = \dfrac{dv}{dt} = 10$ m/s²

(d) $100 + 10t > 500$

   $\Rightarrow 10t > 400$

   $\Rightarrow t > 40$

   After 40 seconds the satellite will break up.

   At this time, the satellite will have travelled

   $((40 \times 100) + (5 \times 40 \times 40)) = (4000 + 8000) = 12\ 000$ metres

A shockwave from an explosion is an expanding sphere.

The volume of the sphere is a function of its radius ... $V = \frac{4}{3}\pi r^3$.

We can calculate the rate at which this volume is expanding as the radius increases in terms of cm^3 per cm increase in radius using the derived function $\frac{dV}{dr} = 4\pi r^2$.

Example 2

A spherical ball is being filled from a compressor.
(a) What radius is it when the volume is 5 litres (5000 cm^3)?
(b) What is the rate of change of the volume per cm of radius when the radius is 10 cm?

Response

(a) $V = \frac{4}{3}\pi r^3 \Rightarrow r = \sqrt[3]{\frac{3V}{4\pi}} = \sqrt[3]{\frac{1500}{4\pi}} = 10\cdot6$cm (to 1.d.p)

(b) $\frac{dV}{dr} = 4\pi r^2$

When $r = 10$, $\frac{dV}{dr} = 400\pi = 1257$ (to nearest whole number).

When the radius is 10 cm the volume of the ball is growing at a rate of 1257 cm^3 per cm of radius.

- Find the gradient of the tangent to the curve $y = f(x)$ at $x = a$.

Example

The point (0, 11) lies on the curve $y = x^3 + 6x^2 + 12x + 11$.
Calculate the gradient of the tangent to the curve at this point.

$f'(x) = \frac{dy}{dx} = 3x^2 + 12x + 12$

At (0, 11) $x = 0$.
$f'(0) = 0 + 0 + 12 = 12$
The tangent to the curve at the point (0, 11) has a gradient of 12.

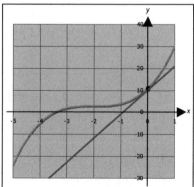

- Find the points on a curve at which the gradient has a given value.

Example

The curve $y = x^3 + 6x^2 + 12x + 11$ has a tangent of gradient 12 at (0, 11).
Where else does it have a tangent of gradient 12?

$f'(x) = 3x^2 + 12x + 12$
$3x^2 + 12x + 12 = 12$
$\Rightarrow 3x^2 + 12x = 0$
$\Rightarrow 3x(x + 4) = 0$
$\Rightarrow x = 0$ or -4
When $x = -4$,
$y = (-4)^3 + 6(-4)^2 + 12(-4) + 11 = -5$.
At $(-4, -5)$ the tangent again has a
gradient of 12.

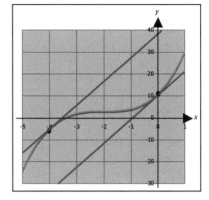

- Apply the facts that, in a given interval:
  if $f'(x) > 0$ then the function f is strictly increasing in that interval;
  if $f'(x) < 0$ then the function f is strictly decreasing in that interval.

  Example
  Prove that $f(x) = \dfrac{x^3}{3} + x^2 + 2x + 1$ is a strictly increasing function.

  $f'(x) = x^2 + 2x + 2$
  We have to show that this is always positive so the best plan is to complete
  the square $f'(x) = (x + 1)^2 + 1$.
  The perfect square can never be less than zero. Given that the derivative can
  never be less than 1.
  So $f'(x) > 0$ for all x and the function f is strictly increasing.

- Apply the fact that if $f'(a) = 0$, then the function f has a stationary value,
  $f(a)$, at $x = a$
  ... $(a, f(a))$ is a stationary point.
  When you apply this fact you should quote it or risk losing the marks.
  A common error is to differentiate a function and then automatically set it
  to zero as if the derivative is always zero. You should always state why you
  are setting it to zero: 'At stationary points $f'(x) = 0$ ...' – then continue.

- Find the stationary points on a curve and determine their nature.

- Sketch a curve with given equation by finding:
  (i)   stationary point(s) and their nature;
  (ii) intersections with the axes;
  (iii) behaviour of y for large positive and negative values of x.

  Example
  Sketch the curve $y = (x - 1)(x^2 - 16)$.

Response

x-intercepts occur when $y = 0$.

$(x - 1)(x^2 - 16) = 0 \Rightarrow x = 1$ or $x = \pm 4$

y-intercepts occur when $x = 0$.

$y = (0 - 1)(0^2 - 16) = 16$

Stationary points occur where $dy/dx = 0$.

At stationary points

$3x^2 - 2x - 16 = 0$

$(3x - 8)(x + 2) = 0$

$x = \frac{8}{3}$ or -2 with corresponding y-values -14.8 (approx) or 36.

This, of course requires that you have covered differential calculus.

A table of signs should accompany this to establish the nature of the stationary points.

x	$\rightarrow$	-2	$\rightarrow$	$\frac{8}{3}$	$\rightarrow$
$3x - 8$	$-$	$-$	$-$	0	$+$
$x + 2$	$-$	0	$+$	$+$	$+$
dy/dx	$+$	0	$-$	0	$+$
gradient	/	—	\	—	/
nature		maximum		minimum	

As x becomes very large and positive both factors of $y = (x - 1)(x^2 - 16)$ become very large and positive and so y becomes very large and positive ... the right-hand tail of the curve will be in the 1st quadrant.

As x becomes very large and negative $(x - 1)$ becomes very large and negative but $(x^2 - 16)$ becomes very large and positive and so y becomes very large and negative ... the left-hand tail of the curve will be in the 3rd quadrant.

Plot this information in a sketch ...

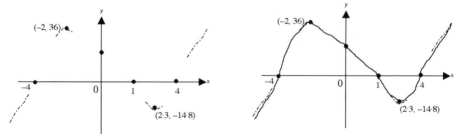

... and join up the clues as smoothly as you can.

- Determine the greatest/least values of a function on a given interval.
  At Higher level, when a function is defined over a closed interval, the maximum/minimum values of the function will occur either at an end point or at a stationary point. You should make sure you examine the function values at each of these points and select the greatest/smallest. It is a common error to examine stationary points only.

  Example

  Find the maximum and minimum vales of the function $f(x) = x^3 - 3x^2 - 9x + 1$ which is defined for $0 \leq x \leq 4$.

  Examine the end-points:
  $f(0) = 0^3 - 3.0^2 - 9.0 + 1 = 1$
  $f(4) = 4^3 - 3.4^2 - 9.4 + 1 = -19$

  Examine the stationary points: $f'(x) = 3x^2 - 6x - 9 = 3(x + 1)(x - 3) = 0$
  So $x = -1$ or 3 at stationary points
  $x = -1$ is outside the domain of the function
  $f(3) = 3^3 - 3.3^2 - 9.3 + 1 = -26$

  By inspection we see that, in this domain the function has a greatest value of 1 (at the end-point where $x = 0$) and has a smallest value of -26 (at the stationary point where $x = 3$).

- Solve optimisation problems using calculus.

  Within particular contexts the ability to find maxima and minima of functions is crucial to solving problems, especially where you are looking to optimise some commodity or course of action.

  Example

  A 12-metre rope is used to cordon off a rectangular area against a wall. How far from the wall should you go to maximise the area cordoned off?

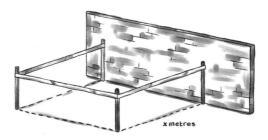

x metres

In plan, the situation looks like this:

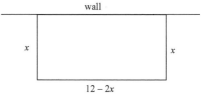

wall

x

x

$12 - 2x$

The area is a function of x: $A(x) = x(12 - 2x) = 12x - 2x^2$; $0 \le x \le 6$.
Differentiating we get $A'(x) = 12 - 4x$.
Stationary points occur when this is zero, i.e. when $x = 3$.
Examine end-points: $A(0) = 0$; $A(6) = 0$.
Examine stationary point: $A(3) = 18$.

The optimum strategy is to take the rope 3 m from the wall to cordon off a maximum area of 18 m^2.

Because we have a closed interval we can be assured that the maximum and minimum values are to be found at end points or stationary points. The largest of these values will automatically be the maximum value of the function. No further examination by, say a table of signs, is needed.

Objective questions

❶ Differentiate $\dfrac{x^3 - x}{x^2}$

   A $1 - x^2$ B $1 - x^{-2}$ C $1 + x^{-2}$ D $\frac{1}{2} x^2 - 1$

   Rough working

   $y = \dfrac{x^3 - x}{x^2} = x^{3-2} - x^{1-2} = x - x^{-1}$

   $\Rightarrow \dfrac{dy}{dx} = 1 - (-1)x^{-1-1} = 1 + x^{-2}$

   Choose option C

❷ $g(x) = x + \sqrt{x}$; $x \ge 0$
   Which of the following describes the behaviour of g when $x = 3$?
   A   increasing
   B   decreasing
   C   stationary at a turning point
   D   stationary at a point of inflexion.

   Rough working

   $g'(x) = 1 + \frac{1}{2}x^{-\frac{1}{2}} = 1 + \dfrac{1}{2\sqrt{x}}$

   $g'(3) = 1 + \dfrac{1}{2\sqrt{3}} > 0$

   Thus the function is *increasing* at $x = 3$.

   When trying to examine the value of the derivative at a particular point, it is better to simplify the expression to avoid negative powers.

   Choose option A

❸ The sketch shows the
function
$f(x) = x^3 - 3x^2 - 9x + 1$ and
its tangent at (1, −10). Find
the equation of the tangent.

A $y - 12x = 2$
B $y + 12x = 2$
C $y + 9x = 1$
D $y + 9x = -1$

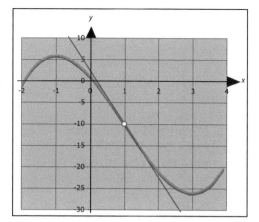

Rough working

$f'(x) = 3x^2 - 6x - 9$
Gradient at $x = 1$:
$3 \times 12 - 6 \times 1 - 9 = -12$.
Equation of tangent:
$y - (-10) = -12(x - 1)$
$\Rightarrow y + 10 = -12x + 12$
$\Rightarrow y + 12x = 2$.

Choose option B

This can be partially checked using the sketch:
... the gradient is negative (eliminating A)
.... the *y*-intercept is greater than 0 (eliminating D).

❹ A particle moves in such a way that its displacement from the origin,
f units, is a function of the time, *t* seconds, measured from the start of
observations.
It is known that $f(t) = t^3 - 5t^2$.
What is the velocity of the particle after 2 seconds?

A   8 units/sec towards the origin
B   8 units/sec away from the origin
C   12 units/sec towards the origin
D   12 units/sec away from the origin

Rough working

$v = f'(t) = 3t^2 - 10t$.
$v_{x=2} = f'(2) = 3.4 - 10.2 = -8$
A positive velocity moves away from the origin: a negative one towards.

Choose option A

5 Where does the graph of the function $f(x) = 2x^3 - 9x^2 + 42x + 1$ have a gradient of 30?

A $x = 1$
B $x = 2$
C $x = 1$ and $x = 2$
D never.

Rough working

$f'(x) = 6x^2 - 18x + 42$
$6x^2 - 18x + 42 = 30 \Rightarrow 6x^2 - 18x + 12 = 0$
$\Rightarrow x^2 - 3x + 2 = 0 \Rightarrow (x - 1)(x - 2) = 0$
$\Rightarrow x = 1$ or $x = 2$

Choose option C

6 A function f is defined as $f(x) = x + \frac{1}{x}; \frac{1}{4} \leq x \leq 3$
What is its maximum value?

A $4\frac{1}{4}$
B $3\frac{1}{3}$
C 2
D $2\frac{1}{2}$

Rough working

$f'(x) = 1 - x^{-2}$
$f'(x) = 0$ at stationary points $\Rightarrow x^2 = 1 \Rightarrow x = 1$ (within given range)
Consider value at S.P. ... $f(1) = 2$.
Consider value at end-points ... $f(\frac{1}{4}) = 4\frac{1}{4}$; $f(3) = 3\frac{1}{3}$.

Maximum value must be one of these three ... $4\frac{1}{4}$ by inspection.

Choose option A

Note that you don't assume the stationary point is going to be a maximum turning point just because the question mentions 'maximum'.

An extended response question that doesn't need a calculator

7 A curve has an equation $y = x + \frac{6}{\sqrt{x}}$, $x > 0$.

Find the equation of the tangent to the curve at the point where $x = 9$.

6 marks

 Response

When $x = 9$, $y = 9 + \frac{6}{3} = 11$

$y = x + 6x^{-\frac{1}{2}}$

$\Rightarrow \dfrac{dy}{dx} = 1 - 3x^{-\frac{3}{2}} = 1 - \dfrac{3}{(\sqrt{x})^3}$

$x = 9 \Rightarrow \dfrac{dy}{dx} = 1 - \dfrac{3}{(\sqrt{9})^3} = \dfrac{8}{9}$

the gradient of the tangent = 8/9.

Equation of tangent: $(y - 11) = \frac{8}{9}(x - 9)$. ────────────⊂▭

Marking scheme

●^{1}	interpret equation to find y	ic
●^{2}	select suitable form to differentiate	ss
●^{3}	select differentiation to find gradient	ss
●^{4}	differentiate negative fractional power	pd
●^{5}	find gradient at $x = 9$	ss
●^{6}	write down equation of tangent	ic

An extended response question that needs a calculator

❽ The cost £C of a construction job depends on the number of workers, x, that the firm employs. They are related by the formula:

$C(x) = \dfrac{600x^2 + 10\,000}{x}$

Find the number of workers that are required to keep the cost to a minimum. Find also the minimum cost. *6 marks*

Response

$C(x) = \dfrac{600x^2 + 10\,000}{x} = 600x + 10\,000x^{-1}$

$C'(x) = 600 - 10\,000x^{-2} = 600 - \dfrac{10\,000}{x^2}$

At a stationary point $C'(x) = 0$

$\Rightarrow x = \sqrt{\dfrac{10\,000}{600}} = 4$ to the nearest whole number.

Set up a table of signs to justify that this is indeed a minimum.

x	$\rightarrow$	$\sqrt{(10\,000/600)}$	$\rightarrow$
$C'(x)$	–	0	+
gradient	\	——	/

The minimum cost = $C(4) = (600 \times 4^2 + 10\,000)/4 = 4\,900$ to the nearest whole number.

The firm should employ 4 workers and the job will be minimised at £4 900.

Marking scheme

- $\bullet^1$ know to differentiate   *ss*
- $\bullet^2$ differentiate   *pd*
- $\bullet^3$ identify crucial aspect viz. $C'(x) = 0$ *ss*
- $\bullet^4$ solve for x *pd*
- $\bullet^5$ justify that this is a minimum   *ic*
- $\bullet^6$ evaluate C at this value of x *pd*

An extended response question that is calculator-neutral

⑨   The sketch is of the curve $y = f(x)$.

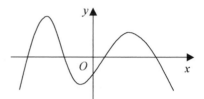

Use it to produce a sketch of $y = f'(x)$ over the same interval.   *3 marks*

(Response)

Make a sketch of the given curve and mark on it where
$f'(x) > 0$, $f'(x) = 0$, $f'(x) < 0$.

Draw a second set of axes below the
first. Transfer the zeroes (the stationary
points) to the x-axis.

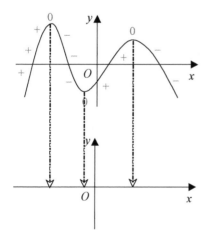

Now connect these three points by
a curve which is above the axis at
the 'plusses' and below it at the
'minuses'.

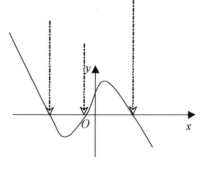

stage 1   stage 2

Marking scheme

- •^{1} identify crucial aspect i.e. $f'(x) = 0$ *ss*
- •^{2} interpret $y = f(x)$ *ic*
- •^{3} sketch a graph of related function   *ic*

Topic Tips

Looking at various marking schemes you can see how often $f'(x)$ is set to the value 0 (zero).

Be explicit about showing this. There is often a mark for **knowing** to differentiate, i.e. for writing '$f'(x) =$' and a separate mark for stating that it equals zero at a stationary point ...

two marks for writing '$f'(x) = 0$ at stationary points'. Either of these marks can go missing if you don't concentrate.

Remember that $f'(x)$ is NOT always zero ... 'at stationary points' is required.

7 Recurrence Relations

What you should know

Related problems

Objective questions

Short response questions that don't need a calculator

An extended response question that needs a calculator

An extended response question that doesn't need a calculator

What you should know

You are expected to know the following facts.

- Know the meaning of the terms:
  sequence – a list of numbers which can be numbered : 1st, 2nd, 3rd, 4th etc. i.e. they are ordered;
  term – each member of the list is referred to as a term, each term being identified by its position in the list;
  ***n*th term** – the general term in the sequence is referred to as the nth term;
  u_n notation – we usually denote the *n*th term by the notation u_n ... u_1 is the 1st term. If a formula for the nth term is found then any term in the sequence can be calculated. A term is expressed as a function of its position *n*, e.g. $u_n = 2n - 1$ would generate the sequence 1, 3, 5, 7, ... as n is successively replaced by 1, 2, 3, 4, ...;
  limit – as *n* tends to infinity there are sequences whose n^{th} term approaches a particular value, forever getting closer to it without arriving.
  e.g. 1, $\frac{1}{2}$, $\frac{1}{4}$, $\frac{1}{8}$, $\frac{1}{16}$, $\frac{1}{32}$, $\frac{1}{64}$, ... has a limit of 0.
  e.g. 0, $\frac{1}{2}$, $\frac{2}{3}$, $\frac{3}{4}$, $\frac{4}{5}$, $\frac{5}{6}$, $\frac{6}{7}$, ... has a limit of 1.
  e.g. $\frac{1}{3}$, $\frac{2}{5}$, $\frac{3}{7}$, $\frac{4}{9}$, $\frac{5}{11}$, $\frac{6}{13}$, ... has a limit of $\frac{1}{2}$.
  e.g. $\frac{3}{2}$, $\frac{6}{3}$, $\frac{9}{4}$, $\frac{12}{5}$, $\frac{15}{6}$, $\frac{18}{7}$, $\frac{21}{8}$, $\frac{24}{9}$, $\frac{27}{10}$, $\frac{30}{11}$, ... has a limit of 3.

- **Recurrence relations** – as well as defining a sequence by defining the nth term as a function of n, we can define the $(n + 1)^{th}$ term as a function of the term before it, i.e. the nth term. We would also have to specify the value of at least one term.

  e.g. $u_1 = 4$ and $u_{n+1} = 2u_n + 1$ would generate the sequence 4, 9, 19, 39, 79, ...

- A recurrence relation of the form $u_{n+1} = mu_n + c$ (m, c constants) is a linear recurrence relation.

- When the multiplier, m, of u_n in the function lies between 1 and -1 a limit of the sequence exists.

- This limit, L, (where it exists) can be found by considering that 'at' the limit, $u_{n+1} = u_n = L$.

  e.g. consider $u_{n+1} = 0.5u_n + 4$.

  A limit exists because we know $-1 < 0.5 < 1$.

  Now let $u_{n+1} = u_n = L$. $L = 0.5L + 4$

  $\Rightarrow 0.5L = 4$

  $\Rightarrow L = 8$

  It is important to establish that the limit exists as part of your answer to finding it. The technique above will give you an answer even when the limit doesn't exist. For example, consider $u_{n+1} = 2u_n + 4$. The technique will return $L = -4$ even though 2 is greater than 1. All it is finding is what is known as a fixed point ... a point where, if you start there, you stay there.

> ⚠ You must say why you know a limit exists or risk losing a mark.

- Where the sequence occurs in some real-life modelling, you should be able to interpret the limit in the context of the mathematical model.

Related problems

You should be able to use the above to solve problems, including:
- forming a recurrence relation to model a given situation;
- deciding if a limit exists;
- finding the limit where it exists;
- interpreting the limit within the context of the question.

Example

A lawn is designed so that 10% of the water drains away in an hour. During a rain shower 10 litres of water land on the lawn in an hour.

(a) Form a recurrence relation to model the situation.

(b) What would happen in the long term if the shower persisted?

(Response)

(a) $u_{n+1} = 0.9u_n + 10$ ... if 10% drains away 90% is left.
(b) $-1 < 0.9 < 1$ so a limit, L, exists.
At the limit, $L = 0.9L + 10 \Rightarrow 0.1L = 10 \Rightarrow L = 100$.
If the shower persists the lawn will get waterlogged with 100 litres of
water.

- Finding conditions to generate specified limits.

Example

In the example above, how efficient must the drainage system be to keep
the limit below 15 litres?

(Response)

$u_{n+1} = au_n + 10$ where $-1 < a < 1$ and a is the fraction left in the lawn after
drainage.
$L = La + 10$
$\Rightarrow L = \dfrac{10}{1-a}$
We wish a limit less than 15.
$\dfrac{10}{1-a} < 15$
$\Rightarrow 10 < 15 - 15a$
$\Rightarrow 15a < 5$
$\Rightarrow a < \dfrac{1}{3}$
In other words a drainage system where more than 66⅔% drains away is
needed.

> Given $u_{n+1} = au_n + b$ then if $-1 < a < 1$ a limit exists.
> The converse is also true ... if you are told a limit exists then you know
> $-1 < a < 1$.

Example

A recurrence relation of the form $u_{n+1} = au_n - 3a + 2$ has a limit equal to 2a.
What is the value of a and what is the limit?

(Response)

Since the limit is 2a we have: $2a = 2a^2 - 3a + 2$
$\Rightarrow 2a^2 - 5a + 2 = 0$
$\Rightarrow (2a - 1)(a - 2) = 0$
$\Rightarrow a = \frac{1}{2}$ or 2.

But since a limit exists, $-1 < a < 1$ and so $a = \frac{1}{2}$.   The limit = 2a = 1.

Objective questions

 What can be said about the recurrence relation $u_{n+1} = -0.9u_n + 38$ as n approaches infinity?

    A   It has a limit of 20.
    B   It has a limit of 380.
    C   It has a limit of −42·22.
    D   It does not have a limit.

Rough working

Check that a limit exists: $-1 < -0.9 < 1$ so a limit exists.
$L = -0.9L + 38 \Rightarrow 1.9L = 38 \Rightarrow L = 20$

Choose option A

 The recurrence relation $u_{n+1} = ku_n + 12$ has a limit of 30. What is the value of k?

    A   0·4   B   0·5   C   0·6   D   0·8

Rough working

A limit exists so we know $-1 < k < 1$.
Form an equation: $30 = 30k + 12 \Rightarrow k = 18 \div 30 = 0.6$.

Choose option C

 The second term of the recurrence relation $u_{n+1} = 0.5u_n + 1$ is 2. What is the difference between the first and third terms?

    A   4   B   2   C   1   D   0

Rough working

$u_2 = 0.5u_1 + 1 = 2 \Rightarrow 0.5u_1 = 1 \Rightarrow u_1 = 2$
$u_3 = 0.5u_2 + 1 = 0.5 \times 2 + 1 = 2$
$u_3 - u_1 = 0$

Choose option D

 The recurrence relation $u_{n+1} = ku_n + t$ has a limit of 1.
Here are some statements made concerning k and t.

    (i) $-t < k < t$ (ii) $k = 1 - t$ (iii) $k = 1 + t$ (iv) k and t are unrelated

Which of the statements are true?

    A   (i) and (ii)   B   (i) and (iii)   C   (ii) only   D   (iv) only

Rough working

Examine the limit: $1 = k.1 + t \Rightarrow k = 1 - t$
[(ii) is true, (iii) is false, (iv) is false.]

Examine the condition: $-t < k < t$.
Since $t = 1 - k$, we have $-1 + k < k < 1 - k \Rightarrow -1 < 0 < 1 - 2k$ [subtracting k throughout].
The right-hand side says $1 - 2k > 0$ for all k ... but if we take a value of $k > 0.5$ it is false.
[(i) is false]. So only (ii) is true.

Choose option C

This topic lends itself to both short response and extended response questions.

Short response questions that don't need a calculator

5 $u_{n+1} = 0.1u_n + 6$ and $v_{n+1} = av_n + 10$ are recurrence relations which have the same limit. Calculate the value of a.   *3 marks*

(Response)

We are told that the relations have a limit, so we know the multipliers lie between 1 and -1.
The limit of the first relation satisfies the equation:
$L = 0.1L + 6 \Rightarrow L = \frac{6}{0.9} = \frac{20}{3}$.

We do not need to simplify this fraction at the moment if we don't want to, as it's mid-way through a calculation. We definitely do not turn it into a decimal such as 6·67 as this is an approximation. In non-calculator papers don't expect lengthy, time-consuming calculations to be a feature.

Use this limit in the second relation: $\frac{20}{3} = a \times \frac{20}{3} + 10$.
Multiply throughout by $\frac{3}{20}$: $1 = a + \frac{3}{2} \Rightarrow a = -\frac{1}{2}$.

Marking scheme

●¹   interpret mathematical terms and notation   *ic*
      [find 1st limit]
●²   marshal facts and make deductions   *ss*
      [apply limit to second relation]
●³   a continuous piece of processing (3+ credit steps)   *pd*
      [solve equation involving fractions]

6   An astronomer spots a star which is shining brightly but getting dimmer as time goes by. The amount of light coming from the star can be modelled by the recurrence relation $u_{n+1} = \dfrac{8u_n + 15}{10}$ where u_n is the amount of light n days after the star was first sighted.

(a) Show that a limit to the recurrence relation exists. *2 marks*

(b) If the amount of light drops below 7 units then the star will be invisible to the naked eye. Will this happen? *2 marks*

Response

(a) $u_{n+1} = \dfrac{8u_n + 15}{10} = \dfrac{8u_n}{10} + \dfrac{15}{10} = 0{\cdot}8u_n + 1{\cdot}5$

A linear relation and the multiplier of u_n, 0·8 lies between 1 and −1 so a limit exists.

>
> Don't be tempted to just answer that the multiplier is less than 1 so a limit exists.
> This could lead to claims that a limit exists when the multiplier is −3 for example.
> You should also declare that it is greater than −1.

(b) The limit, L, satisfies: $L = 0{\cdot}8L + 1{\cdot}5 \Rightarrow 0{\cdot}2L = 1{\cdot}5 \Rightarrow L = 7{\cdot}5$.
This is greater than 7 so you should always be able to see the star.

Marking scheme

$\bullet^1$ select a suitable form in which to rewrite expression *ss*
   [i.e. linear ... to identify multiplier]

$\bullet^2$ marshal facts and make deductions *ss*
   [limit exists]

$\bullet^3$ marshal facts and make deductions *ss*
   [find limit]

$\bullet^4$ express solution in language appropriate to context *ic*
   [solution in astronomical context]

Care should be taken here. If the context takes 0·8 of u_n and **then** adds 1·5 later, the value of u_n will oscillate between 6 and 7·5 ... perhaps affecting the practicalities of your answer, e.g. the star would blink in and out of sight.

However, in this case, there is no suggestion that the recurrence relation is modelling a two-stage event.

An extended response question that needs a calculator

❼ A small mobile library starts with a stock of 500 books. Because of wear and tear, it discards 28% of its stock at the end of the year. At the same time another 360 new books are added to the stock.

(a) The librarian will be happy when the stock gets above 1000 books. In which year will this happen?   *3 marks*

(b) Write down a recurrence relation for the size of the stock after the addition of the top-up books. Let u_1 represent the size of the stock after the 1st addition.   *1 mark*

(c) There is only room for 1200 books in the mobile library. Will there come a time when the above stock management will become a problem?   *4 marks*

Response

(a) At the start: 500 books
    1st addition: 720 books
    2nd addition: 878 books
    3rd addition: 992 books
    4th addition: 1074 books

So at the end of the fourth year the stock goes above 1000.

> Note that calculators with 'ANS' buttons can reduce the amount of work at this repetitive number crunching:
> Step 1: Enter the initial value ... 500 =
> Step 2: ANS $\times$ 0·72 + 360 = will produce the first answer ... 720.
> Step 3: Each press of the = button will generate the next value: 878·4, 992·448, 1074·56256 ...
> Do not use this method to try to justify the existence of a limit. Special cases, whether there are 3 or 33 of them, will never prove a general statement.

(b) $u_{n+1} = 0.72u_n + 360$; $u_0 = 500$

(c) This recurrence relation has a limit, L, since $-1 < 0.72 < 1$.
$L = 0.72L + 360$
$\Rightarrow 0.28L = 360$
$\Rightarrow L = 1286$ (to nearest whole number)
The current stock management system will lead to overstocking.

Marking scheme

•^{1}	identify multiplier [0·72]	*ic*
•^{2}	process the data four times	*pd*
•^{3}	interpret the findings [at the end of the fourth year]	*ic*
•^{4}	interpret a collection of facts	*ic*
•^{5}	interpret, justifying existence of limit	*ic*
•^{6}	strategy for finding limit [equation]	*ss*
•^{7}	solve the equation	*pd*
•^{8}	communicate in language appropriate to situation	*ic*

An extended response question that doesn't need a calculator

8 An angling club would like to keep its loch well stocked.
The number of fish in the loch can be modelled by a recurrence relation of the form $u_{n+1} = au_n + b$ where a and b are constants and u_n is the number of fish in the loch at the start of year n.

(a) Careful record keeping shows that $u_1 = 1000$, $u_2 = 900$, $u_3 = 800$. Calculate the values of a and b.　　　　　　　　　　　　*3 marks*

(b) It is found that the above system leads to depleted stocks. Steps are taken leading to new values of
$u_1 = 1000$, $u_2 = 950$, $u_3 = 940$.

　(i) What are the values of a and b now?　　　　　　　　*1 mark*

　(ii) What are the long-term prospects of the fish stocks now?　　　　　　　　　　　　　　　　　　　*4 marks*

(Response) ─────────────────────────────

(a) $u_1 = 1000$, $u_2 = 900$ lead to　　$900 = 1000a + b$　...　①
　　$u_2 = 900$, $u_3 = 800$ lead to　　$800 = 900a + b$　　...　②
　　Subtracting gives $100 = 100a \Rightarrow a = 1$.
　　Substituting in ① gives $900 = 1000 + b \Rightarrow b = -100$.

(b) (i)　$u_1 = 1000$, $u_2 = 950$ lead to　$950 = 1000a + b$　...　①
　　　$u_2 = 950$, $u_3 = 940$ lead to　$940 = 950a + b$　　...　②
　　　Subtracting gives $10 = 50a \Rightarrow a = \frac{1}{5}$.
　　　Substituting in ① gives $950 = 1000 \times \frac{1}{5} + b \Rightarrow b = 750$.

　(ii)　The recurrence relation is $u_{n+1} = \frac{1}{5}u_n + 750$.
　　　Since $-1 < \frac{1}{5} < 1$ a limit, L, exists.
　　　$L = \frac{1}{5}L + 750 \Rightarrow \frac{4}{5}L = 750 \Rightarrow L = 938$ to nearest whole number.

　　　The fish population after restock should stabilise at 938 fish.

─────────────────────────────────

Marking scheme

- ●1　interpret a collection of facts before proceeding　　　*ic*
- ●2　select the strategy of a system of equations　　　　　*ss*
- ●3　carry out process with up to three Credit processes　　*pd*
- ●4　imitate the steps of a mathematical argument　　　　*ic*
- ●5　interpret, justifying existence of limit　　　　　　　*ic*
- ●6　strategy for finding limit [equation]　　　　　　　　*ss*
- ●7　solve the equation　　　　　　　　　　　　　　　*pd*
- ●8　communicate in language appropriate to situation　　*ic*

The model only predicts the population of the fish immediately after restock. It can't be used to compute intermediate values. We have no way of telling how the 80% loss that seems to occur every year is happening. Is it just that deaths outstrips births? Is there an angling competition just before restock?

Topic Tips

You are expected to justify the existence of a limit before looking for it. If you miss this step out, you will probably lose a mark.

As stated earlier, the technique for finding the limit will produce a value even where there is no limit.

This is the most commonly lost mark ... and remember one mark can separate an A grade from a B grade.

8 Factor/Remainder Theorem and Quadratic Theory

> *What you should know*
>
> *Related problems*
>
> *Objective questions*
>
> *Extended response questions that don't need a calculator*
>
> *An extended response question that needs a calculator*

What you should know

You are expected to know the following facts.

- Given the equation $f(x) = 0$, h is said to be a root of the equation if $f(h) = 0$. h is referred to as a *zero* of the function f.

- The Remainder Theorem
  ... When a polynomial, $f(x)$, is divided by $(x - h)$ the remainder will be $f(h)$.

- The Factor Theorem
  ... If $f(h) = 0$ then $(x - h)$ is a factor of the polynomial $f(x)$.

- The converse of the Factor Theorem is true, i.e. if $(x - h)$ is a factor of the polynomial $f(x)$ then $f(h) = 0$.

- If polynomial $f(x)$ is of the third or higher degree, then $f(x)$ can be expressed as a product of quadratic and linear factors.
  [For the exam, when asked to factorise, at most one of the factors will be quadratic and the rest will be linear.]

- Know that the roots of $ax^2 + bx + c = 0$, $a \neq 0$, are $x = \dfrac{-b \pm \sqrt{b^2 - 4ac}}{2a}$.

- Know that the discriminant of $ax^2 + bx + c = 0$ is $b^2 - 4ac$.

- Know that the roots of a quadratic equation are:
  (i) real when $b^2 - 4ac \geq 0$;
  (ii) non-real when $b^2 - 4ac < 0$;
  (iii) equal when $b^2 - 4ac = 0$;
  (iv) distinct when $b^2 - 4ac > 0$.

  These conditions are important for exam purposes ... often if the relevant condition is not quoted, marks will not be given.

  Note that, although not essential for exam purposes, it is useful to know that when the discriminant is a perfect square the roots are rational ... and the quadratic expression will factorise. It can waste a lot of time looking for factors that aren't there.

- Know the condition for tangency/intersection of a straight line and a parabola, i.e. consider the line $y = px + q$ and the parabola $y = ax^2 + bx + c$. Their relationship can be discovered by exploring the roots of the equation $px + q = ax^2 + bx + c$.
  (i) Distinct real roots mean the line cuts the parabola in two places.
  (ii) Equal roots tells us that the line is a tangent to the parabola.
  (iii) Non-real roots tell us the line does not cut the parabola.

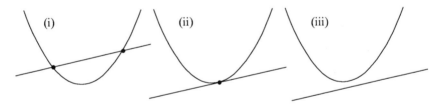

- Know how to solve quadratic inequalities, $ax^2 + bx + c \geq 0$ or $ax^2 + bx + c \leq 0$. It is quite possible to solve these inequalities analytically but at Higher level it is safer to solve the corresponding equation, sketch the curve and figure out the solution from this.

  Example   Find where $2x^2 + 7x + 3 \geq 0$.

Response

First solve $2x^2 + 7x + 3 = 0$.
Factorising: $(2x + 1)(x + 3) = 0$
$$\Rightarrow x = -\tfrac{1}{2} \text{ or } x = -3.$$
A quick sketch gives:

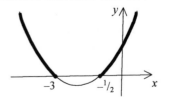

The part of the curve where $y \geq 0$ is highlighted.
$y \geq 0$ where $x \geq -\tfrac{1}{2}$ or where $x \leq -3$

- Know how to form a quadratic equation with given roots.
- Know how to prove that an equation has a root between two given values.
- Know how to find that root to a required degree of accuracy.

Related problems

The candidate should be able to use the above information to solve the following kinds of problems.

- Using the Remainder Theorem to find the remainder on dividing a polynomial $f(x)$ by $x - h$.

  Example

  When $f(x) = x^3 + 2x^2 - 3x + 4$ is divided by $(x - 2)$ the remainder will be $f(2)$. There are two main ways to perform this division ... select the one you are familiar with.

  (i) *direct division*

$$
\begin{array}{r}
x^2 + 4x + 5 \\
x - 2 \enclose{longdiv}{x^3 + 2x^2 - 3x + 4} \\
\underline{x^3 - 2x^2} \\
4x^2 - 3x + 4 \\
\underline{4x^2 - 8x} \\
5x + 4 \\
\underline{5x - 10} \\
14
\end{array}
$$

  So, $(x^3 + 2x^2 - 3x + 4) \div (x - 2) = x^2 + 4x + 5$ remainder 14.
  Check that $f(2) = 2^3 + 2.2^2 - 3.2 + 4 = 8 + 8 - 6 + 4 = 14$.

  (ii) *synthetic division* $(x^3 + 2x^2 - 3x + 4) \div (x - 2)$

$$
\begin{array}{r|rrrr}
2 & 1 & 2 & -3 & 4 \\
  &   & 2 & 8 & 10 \\
\hline
  & 1 & 4 & 5 & \mathbf{14}
\end{array}
$$
     ... coeffs of the cubic

     ... coeffs of quad^c and **f(2)**

  This method is quicker but it's easy to get it wrong if you have a bad memory.

- Use the Factor Theorem to determine the factors of a polynomial.
  [At most one will be an irreducible quadratic and the remainder will be linear.]
  e.g. Level C $f(x) = (x - 2)(x - 3)(x^2 + x + 6)$.
              i.e. no need to consider $h \in Q$ for factor $x - h$.
  Level A/B $f(x) = (2x - 1)(3x + 2)(2x - 5)$.

  By the Remainder Theorem, when we divide a polynomial, $f(x)$ by $(x - h)$ we get a remainder of $f(h)$. If this is zero and there is no remainder then $(x - h)$ must be a factor of $f(x)$ ... with the result of the division being the complementary factor.

Example

Factorise $f(x) = x^3 + 5x^2 + 2x - 8$.

Because the constant term is –8 we should consider $x = \pm1, \pm2, \pm4$ and ±8 and evaluate $f(x)$ at these values … starting with the simplest.

$$
\begin{array}{r|rrrr}
1 & 1 & 5 & 2 & -8 \\
  &   & 1 & 6 & 8 \\
\hline
  & 1 & 6 & 8 & 0
\end{array}
$$
… lucky first time.

The fact that the remainder is zero should be highlighted in your answer … this will exhibit your knowledge of the Remainder and Factor Theorems. A communication mark may go missing if you don't. You should now specify the two factors you've found …

The remainder is zero so $x - 1$ is a factor of $f(x)$

$f(x) = (x - 1)(x^2 + 6x + 8)$

… again a communications mark is at stake.

Now factorise the polynomial that is left … in this case a quadratic …

$f(x) = (x - 1)(x + 2)(x + 4)$.

- Determine the roots of a polynomial equation.

C level Example

Solve $2x^4 - 5x^3 - 11x^2 + 20x + 12 = 0$.

First factorise by considering the factors of 12.

$$
\begin{array}{r|rrrrr}
1 & 2 & -5 & -11 & 20 & 12 \\
  &   & 2 & -3 & -14 & 6 \\
\hline
  & 2 & -3 & -14 & 6 & 18
\end{array}
\qquad
\begin{array}{r|rrrrr}
-1 & 2 & -5 & -11 & 20 & 12 \\
   &   & -2 & 7 & 4 & -24 \\
\hline
   & 2 & -7 & -4 & 24 & -12
\end{array}
$$

$$
\begin{array}{r|rrrrr}
2 & 2 & -5 & -11 & 20 & 12 \\
  &   & 4 & -2 & -26 & -12 \\
\hline
  & 2 & -1 & -13 & -6 & 0
\end{array}
$$

The remainder is zero so $x - 2$ is a factor … [this is an important *ic* line]

$f(x) = (x - 2)(2x^3 - x^2 - 13x - 6)$.

Now factorise the cubic … only examining the factors of 12 not yet considered …

$$
\begin{array}{r|rrrr}
-2 & 2 & -1 & -13 & -6 \\
   &   & -4 & 10 & -6 \\
\hline
   & 2 & -5 & -3 & 0
\end{array}
$$

The remainder is zero so $x - (-2)$ is a factor …i.e. $x + 2$

$f(x) = (x - 2)(x + 2)(2x^2 - 5x - 3)$.

Finally factorise the quadratic if possible ... $b^2 - 4ac > 0$ (see below)
$f(x) = (x - 2)(x + 2)(x - 3)(2x + 1)$.
Don't forget to solve the equation!
$(x - 2)(x + 2)(x - 3)(2x + 1) = 0$
$\Rightarrow (x - 2) = 0$ or $(x + 2) = 0$ or $(x - 3) = 0$ or $(2x + 1) = 0$
$\Rightarrow x = 2$ or $x = -2$ or $x = 3$ or $x = -\frac{1}{2}$.

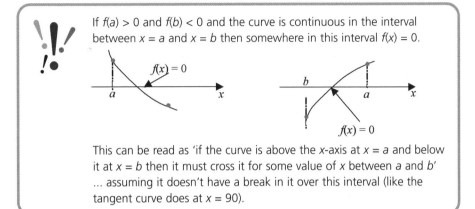

If $f(a) > 0$ and $f(b) < 0$ and the curve is continuous in the interval between $x = a$ and $x = b$ then somewhere in this interval $f(x) = 0$.

This can be read as 'if the curve is above the x-axis at $x = a$ and below it at $x = b$ then it must cross it for some value of x between a and b' ... assuming it doesn't have a break in it over this interval (like the tangent curve does at $x = 90$).

If you are aiming for an A or B, you will need to be able to use this technique. See the following example.

Example, A/B grade
Solve $12x^3 - 8x^2 - 3x + 2 = 0$.

An initial hunt for a root should involve a look at $f(-1)$, $f(0)$, and $f(1)$.

1	12	−8	−3	2
		12	4	1
	12	4	1	**3**

−1	12	−8	−3	2
		−12	20	−17
	12	−20	17	**−15**

Note that $f(1)$ is positive and $f(-1)$ is negative ... we can expect a root in the interval $-1 < x < 1$ ... now $f(0) = 2$,
so we can reduce this interval ... $-1 < x < 0$
Examining the cubic term and the constant term we could expect factors such as $(2x + 1)$, $(3x + 1)$, $(4x + 1)$, $(6x + 1)$, $(3x + 2)$ ...
Try $(2x + 1)$ ... if you're using synthetic division $(x + \frac{1}{2})$

−½	12	−8	−3	2
		−6	7	−2
	12	−14	4	**0**

So the expression factorises as $(x + \frac{1}{2})(12x^2 - 14x + 4)$.

Transferring the common factor of 2 from the second factor to the first we get ...

$(2x + 1)(6x^2 - 7x + 2)$ ... and factorizing the quadratic expression gives $(2x + 1)(3x - 2)(2x - 1)$.

Now answer the question:

$12x^3 - 8x^2 - 3x + 2 = 0$

$\Rightarrow (2x + 1)(3x - 2)(2x - 1) = 0$

$\Rightarrow 2x + 1 = 0$ or $3x - 2 = 0$ or $2x - 1 = 0$

$\Rightarrow x = -\frac{1}{2}$ or $\frac{1}{2}$ or $\frac{2}{3}$.

- Use the discriminant to determine the nature of the roots of a quadratic equation.

> Given $ax^2 + bx + c = 0$ then $x = \dfrac{-b \pm \sqrt{b^2 - 4ac}}{2a}$ ... note that $a \neq 0$.

The expression under the square root sign, $b^2 - 4ac$, is called the *discriminant*.

If it is negative then no *real* solutions exist ... since no real number has a negative square root.

If it is zero then the only solution is $x = \dfrac{-b}{2a}$.

(This *one-root* situation is also referred to by the expression *two equal roots* or *two coincident roots*.)

If it is greater than zero then you will get two distinct solutions.

[Note: If the discriminant is a perfect square then the roots will be *rational*. This can be useful to know but is not strictly necessary for the exam.]

To summarise:

$b^2 - 4ac < 0 \Rightarrow$ no real roots

$b^2 - 4ac = 0 \Rightarrow$ one root (or two coincident roots)

$b^2 - 4ac > 0 \Rightarrow$ two distinct real roots

Example

Find the nature of the roots of the equation $2x^2 + 3x - 4 = 0$.

Comparing this equation with the standard equation

$a = 2$, $b = 3$ and $c = -4$

$\Rightarrow b^2 - 4ac = 3^2 - 4.2.(-4) = 9 + 32 = 41$

$\Rightarrow b^2 - 4ac > 0$

The equation has **two distinct, real** roots.

All the emphasised words above are essential to gain the communication marks.

The line which compares the discriminant with zero is also essential for getting a strategy mark.

- Conversely, find the condition that the roots of a quadratic equation have a particular nature.

Example, level C

For what values of k does the equation $x^2 - 3x + k = 0$ have real roots?

$a = 1$, $b = -3$ and $c = k$

$\Rightarrow b^2 - 4ac = (-3)^2 - 4.1.k = 9 - 4k$

For real roots $b^2 - 4ac \geq 0$

$\Rightarrow 9 - 4k \geq 0$

$\Rightarrow 9 \geq 4k$

$\Rightarrow k \leq \frac{9}{4}$

The condition for real roots must be stated ... and note that it must cover the 'greater than' and the 'equal to' cases. These are commonly lost marks.

Example, level A/B

If $\dfrac{(x-3)^2}{x^2+3} = k$, $k \in \mathbf{R}$, find values of k such that the given equation has two coincident roots.

First you must rearrange the equation into the standard quadratic form. Multiply both sides by $x^2 + 3$ ... we can do this because $x^2 + 3 \neq 0$

$\Rightarrow (x-3)^2 = k(x^2+3)$

$\Rightarrow x^2 - 6x + 9 = kx^2 + 3k$

$\Rightarrow (1-k)x^2 + (-6)x + (9-3k) = 0$

$a = (1-k)$, $b = -6$ and $c = (9-3k)$

$\Rightarrow b^2 - 4ac = (-6)^2 - 4.(1-k).(9-3k) = 36 - 36 + 48k - 12k^2$

$\Rightarrow b^2 - 4ac = 12k(4-k)$.

Now state the conditions for equal roots ...

$\Rightarrow b^2 - 4ac = 0$ for equal roots

$\Rightarrow 12k(4-k) = 0$

$\Rightarrow k = 0$ or $k = 4$ for coincident roots.

Notes:
Deciding to rearrange the equation into the standard quadratic form is selecting the strategy. Make it obvious that you are doing so to make sure you get the mark. Once again, you must state the exact conditions the discriminant satisfies, i.e. you must correctly compare it to zero or risk losing a mark.

- Find the condition for tangency or the intersection of a straight line with a parabola.

  The straight line $y = mx + k$ will meet the parabola $y = ax^2 + bx + c$ when the equation $mx + k = ax^2 + bx + c$ has real roots.

  If these roots are coincident then the line is a **tangent** to the parabola.

  If these roots are distinct then the line forms a **chord** of the parabola, cutting the parabola in two places.

  Since the equation is a quadratic, we can use its discriminant to determine the line's relation to the parabola or to determine conditions that will make the line a tangent.

  Example

  A line passes through the point (0, 6). What must its gradient be so that it is a tangent to the parabola $y = x^2 + 2x + 10$?

  The line has an equation of the form
  $y = mx + 6$.
  It cuts the parabola when
  $mx + 6 = x^2 + 2x + 10$
  $\Rightarrow x^2 + (2 - m)x + 4 = 0$.
  For a tangent we want intersection at one point only, i.e. coincident roots.
  Now $b^2 - 4ac = 0$ for equal roots
  $\Rightarrow (2 - m)^2 - 16 = 0$
  $\Rightarrow 2 - m = \pm 4$
  $\Rightarrow m = 2 \pm 4$
  $\Rightarrow m = -2$ or 6
  Both $y = -2x + 6$ and $y = 6x + 6$ are tangents to the parabola passing through (0, 6).

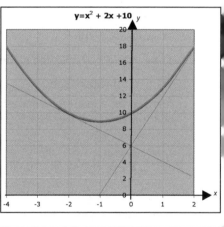

- Solve quadratic inequalities,
  $ax^2 + bx + c \geq 0$ or $ax^2 + bx + c \leq 0$.

  Example, level C

  The diagram shows a sketch of the graph of $y = x^2 + x - 6$.

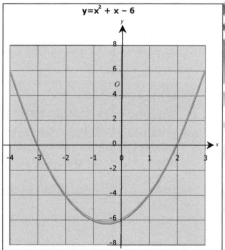

Solve $x^2 + x - 6 > 0$.

By inspection, the function is bigger than zero when it is above the x-axis.
This happens when $x < -3$ **or** $x > 2$.
*Don't be tempted to write 'and' in place of 'or' ... there are no values of x
which fit the description x < -3 **and** x > 2.*
*Also note that the question asked for when the function was bigger than,
not bigger than or equal to zero.*

Example, level A/B

Find the real values of x
satisfying $x^2 + 3x \leq 4$.
Start by standardizing the form:
$x^2 + 3x - 4 \leq 0$.
*Solve the corresponding
equation:*
$x^2 + 3x - 4 = 0$
$\Rightarrow (x - 1)(x + 4) = 0$
$\Rightarrow x = 1$ or -4.
Use this to help you sketch the
curve.

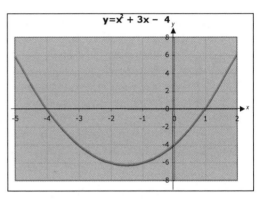

$y = x^2 + 3x - 4$

By inspection, the function is less than or equal to zero when it is on or
below the x-axis.
This happens in the interval $-4 \leq x \leq 1$.
*Note that in the A/B question, the decision to draw the necessary sketch was
yours.*

- Form a quadratic equation with given roots.
  *If the roots of a quadratic equation are a and b then the quadratic equation
  has the form (x − a)(x − b) = 0.*
  *If the question asks for a quadratic function with zeros a and b then it will
  have the form f(x) = k(x − a)(x − b). To establish the value of k, the value of the
  function must be known for some other value of x apart from x = a or x = b.*

  Example, level C

  Find an equation of the form $x^2 + bx + c = 0$ which has roots
  $x = 1$ and $x = 5$.
  The equation will be $(x - 1)(x - 5) = 0$.
  Expanding the brackets gives $x^2 - 6x + 5 = 0$.

Example, level A/B

Find a quadratic function of the form $f(x) = ax^2 + bx + c$ which has zeros $x = 1$ and $x = 5$ and which passes through the point (0, 10).

The equation will be of the form $f(x) = k(x - 1)(x - 5)$.
Expanding the brackets gives $f(x) = kx^2 - 6kx + 5k$.
We know $f(0) = 10 \Rightarrow 10 = 0 - 0 + 5k \Rightarrow k = 2$.
So the required function is: $f(x) = 2x^2 - 12x + 10$.

● Prove that an equation has a root between two given values.

> If you have a calculator with an [ANS] facility (a button which you can press to recall the last answer) and you wish to evaluate, say, $x^3 - 2x^2 - 1$ at $x = 2 \cdot 12223$:
>
> Step 1: type 2·12223 = ... this puts 2·12223 into the memory;
> Step 2: type $[ANS]^3 - 2[ANS]^2 - 1 =$
> ... to get the answer −0·449493171.

Example

(a) Show that the equation $x^3 - 2x^2 - 1 = 0$ has a root between 2 and 3.
(b) Find this root correct to 2 decimal places.

$$f(2) = -1 \text{ (using the calculator)} \qquad ... f(2) < 0$$
$$f(3) = 8 \qquad\qquad\qquad\qquad\qquad ... f(3) > 0$$

So there exists x such that $2 < x < 3$ and $f(x) = 0$.
$f(2)$ is closer to zero that $f(3)$ ... so perhaps examine $f(2 \cdot 2)$
$f(2 \cdot 2) = -0 \cdot 032$ ... root now lies between $2 \cdot 2 < x < 3$
$f(2 \cdot 3) = 0 \cdot 587$ ... root now lies between $2 \cdot 2 < x < 2 \cdot 3$
$f(2 \cdot 21) = 0 \cdot 025661$ ... root now lies between $2 \cdot 2 < x < 2 \cdot 21$
$f(2 \cdot 205) = -0 \cdot 00328$ ... root now lies between $2 \cdot 205 < x < 2 \cdot 21$
$f(2 \cdot 206) = 0 \cdot 00248$ ... root now lies between $2 \cdot 205 < x < 2 \cdot 206$
Corrected to 2 decimal places 2·205 and 2·206 are both 2·21
$x = 2 \cdot 21$ correct to 2 d.p. is the approximate solution required.

Note that in selecting each guess we took guidance from the fact that we knew the answer was between two numbers and **probably** closer to the one that produced a value closer to zero. When neither seemed particularly close we chose the half-way mark. For a communication mark, it is essential that you state why you stopped where you did.

Objective questions

❶ Which of the options is a factor of $x^4 - 4x^3 + x^2 + 6x$?

   A $(x - 1)$ B $(x + 2)$ C $(x - 2)$ D $(x + 3)$

Rough working

$f(1) = 1 - 4 + 1 + 6 = 4 \neq 0$
$f(-2) = 16 + 32 + 4 - 12 = 40 \neq 0$
$f(2) = 16 - 32 + 4 + 12 = 0$ ... $x - 2$ is a factor

Choose option C

Confirm your answer by checking that the untried option doesn't work.
$f(-3) = 81 + 108 + 9 - 18 = 180 \neq 0$

❷ The polynomial $f(x) = x^3 + 2x + m$ has a zero at $x = 1$.
What are the factors of $f(x)$?

   A $(x - 1)(x + 1)(x + 3)$
   B $(x - 1)(x + 1)(x + 2)$
   C $(x + 1)(x^2 + x + 3)$
   D $(x - 1)(x^2 + x + 3)$

Rough working

$f(1) = 1^3 + 2.1 + m = 0 \Rightarrow m = -3$
$f(x) = x^3 + 2x - 3$ has a zero at $x = 1$

```
1 | 1   0   2  -3
  |     1   1   3
  -----------------
    1   1   3   0        ... remainder zero ...  (x − 1) is a factor
```

$x^2 + x + 3$ is also a factor ... discriminant is $1 - 12 < 0$ ... irreducible.

Choose option D

Note that when setting up the synthetic division, a zero had to be placed as the coefficient of the x^2 term.

A quick check can be done by multiplying out the chosen option.

❸ What must be true about m so that the equation
$y = x^2 + 4x + m$ has real, distinct roots?
   A $m = 4$ B $m \leq 4$ C $m < 4$ D $m > 4$

Rough working

$b^2 - 4ac > 0$ for real distinct roots
$\Rightarrow 4^2 - 4.1.m > 0$
$\Rightarrow 16 - 4m > 0$
$\Rightarrow m < 4$

Choose option C

Note that *distinct* roots are required so the '... or equal to ...' is wrong.

④ When will the equation $x^2 + (k + 1)x + k = 0$ where k is a real number constant, have no real roots?

A When $k = 1$
B When $k < 1$
C When $k \geq 1$
D Never.

Rough working

$b^2 - 4ac < 0$ for non-real roots
$\Rightarrow (k + 1)^2 - 4.1.k < 0$
$\Rightarrow k^2 + 2k + 1 - 4k < 0$
$\Rightarrow k^2 - 2k + 1 < 0$
$\Rightarrow (k - 1)^2 < 0$

Since $(k - 1)^2$ is a perfect square then this can never happen ... the lowest a square can get to is zero.

Choose option D

⑤ Solve the quadratic inequality $x^2 + 2x - 15 > 0$.

A $x > 3$ or $x < -5$
B $-5 < x < 3$
C $x > 5$ or $x < -3$
D $-3 < x < 5$

Rough working

Solve the corresponding equation: $x^2 + 2x - 15 = 0 \Rightarrow (x - 3)(x + 5) = 0$
$\Rightarrow x = 3$ or -5.

A quick sketch

... to help us identify when the curve is above the x-axis,

... when $x > 3$ or when $x < -5$.

Choose option A

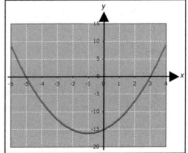

6 Find a quadratic equation where one root is double the other, where the roots are both whole numbers and where one of the roots is 3.

A $2x^2 - 9x + 9 = 0$
B $x^2 - 9x + 18 = 0$
C $x^2 - 9x + 9 = 0$
D $2x^2 - 9x + 18 = 0$

Rough working

With one root 3 and the other a whole number ... it must be 6.
The equation is therefore $(x - 3)(x - 6) = 0$.
Expanded this becomes $x^2 - 9x + 18 = 0$.

Choose option B

Note that if the restriction about the roots being whole numbers is lifted, $x = 3$ and $x = \frac{3}{2}$ would work ... $(2x - 3)(x - 3) = 0$ ... $2x^2 - 9x + 9 = 0$.

7 In which of the following regions does a root of the equation $x^3 + 2x^2 - 15x - 12 = 0$ lie?

A $-2 < x < -1$
B $-1 < x < 0$
C $0 < x < 1$
D $1 < x < 2$

Rough working

$f(-2) = (-2)^3 + 2(-2)^2 - 15(-2) - 12 = 18$
$f(-1) = (-1)^3 + 2(-1)^2 - 15(-1) - 12 = 4$
$f(0) = (0)^3 + 2(0)^2 - 15(0) - 12 = -12$
So $f(-1) > 0$ and $f(0) < 0$ ... a root must lie in the region $-1 < x < 0$.

Choose option B

Extended response questions that don't need a calculator

8 (a) $x + 3$ is a factor of the cubic expression $2x^3 + kx^2 - 4x - 3$.
  Find the value of k.   *3 marks*

  (b) For this value of k, solve the equation $2x^3 + kx^2 - 4x - 3 = 0$.   *2 marks*

Response

(a)

-3	2	k	-4	-3
		-6	$-3k + 18$	$9k - 42$
	2	$k - 6$	$-3k + 14$	$9k - 45$

The remainder is $9k - 45$. Since $x + 3$ is a factor, the remainder equals zero. $9k - 45 = 0 \Rightarrow k = 5$

(b) By the above, letting $k = 5$
$$2x^3 + 5x^2 - 4x - 3 = 0$$
$$\Rightarrow (x + 3)(2x^2 - x - 1) = 0$$
$$\Rightarrow (x + 3)(2x + 1)(x - 1) = 0$$
$$\Rightarrow x = -3 \text{ or } -\tfrac{1}{2} \text{ or } 1$$

Marking scheme

Part a)
- $\bullet^1$ identify crucial aspect   *ss*
  [divide by $x + 3$ to get zero remainder]
- $\bullet^2$ perform the division   *pd*
- $\bullet^3$ solve the resultant equation for k *pd*

Part b)
- $\bullet^4$ determine the quadratic factor   *ss*
  [divide by $x + 3$ to get zero remainder]
- $\bullet^5$ solve factorised equation   *pd*

9 $y = 3x^2 - x + 2$ and $y = 1 + 3x - x^2$ are two parabolas.
(a) Form a single quadratic equation that can be used to discover any points of intersection between the two curves.   *2 marks*
(b) By using the discriminant to help, describe how and where the two curves intersect.   *3 marks*

Response

(a) $3x^2 - x + 2 = 1 + 3x - x^2$
$$\Rightarrow 4x^2 - 4x + 1 = 0$$
(b) $b^2 - 4ac = (-4)^2 - 4.4.1 = 16 - 16 = 0$
The discriminant is zero which means the curves intersect at only one point ... $\left(\dfrac{1}{2}, \dfrac{9}{4}\right)$.

Marking scheme

Part a)
- $\bullet^1$ identify crucial aspect   *ss*
  [equating expressions for y]
- $\bullet^2$ translate into another suitable form   *ic*
  [standard quadratic form]

Part b)
- $\bullet^3$ calculate discriminant   *pd*
- $\bullet^4$ interpret conditions as equal roots   *ic*
- $\bullet^5$ determine point of intersection   *ss*

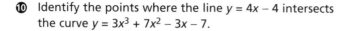

⑩ Identify the points where the line $y = 4x - 4$ intersects the curve $y = 3x^3 + 7x^2 - 3x - 7$.

7 marks

(Response) ────────────────────────────

At the intersections $3x^3 + 7x^2 - 3x - 7 = 4x - 4$

$\Rightarrow 3x^3 + 7x^2 - 7x - 3 = 0$.

Look for one root:

```
1 │  3    7    -7   -3
  │       3    10    3
  └────────────────────
     3   10    3    0        ... remainder zero ...  (x – 1) is a factor.
```

Equation becomes $(x - 1)(3x^2 + 10x + 3) = 0$

$\Rightarrow (x - 1)(x + 3)(3x + 1) = 0$ [... or use the quadratic formula if you can't spot factors]

$\Rightarrow x = 1$ or -3 or $-\frac{1}{3}$.

Points of intersection are $(1, 0)$, $(-3, -16)$ and $(-\frac{1}{3}, 5\frac{1}{3})$

Marking scheme

- •^{1} identify crucial aspect of problem   *ss*
  [equating expressions for *y*]
- •^{2} translate into another suitable form   *ic*
  [standard quadratic form]
- •^{3} strategy: Remainder Theorem to find factor   *ss*
- •^{4} perform division   *pd*
- •^{5} interpret result [zero means factor]   *ic*
- •^{6} process data [get quadratic factor]   *pd*
- •^{7} solve for roots   *pd*

⑪ Find the equation of the cubic function represented by this sketch.

5 marks

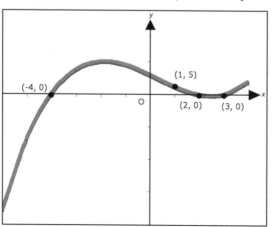

Cubic must be of form $f(x) = k(x + 4)(x - 2)(x - 3)$.
We see that $f(1) = 5 \Rightarrow k(1 + 4)(1 - 2)(1 - 3) = 5$
$\Rightarrow k.5.(-1)(-2) = 5$
$\Rightarrow 10k = 5$
$\Rightarrow k = \frac{1}{2}$
So the function is: $f(x) = \frac{1}{2}(x + 4)(x - 2)(x - 3)$.

Marking scheme

- $\bullet^1$ identify crucial aspect of problem   ss
  [identify factors from zeros]
- $\bullet^2$ interpret sketch   ic
  [identify zeros and $f(1) = 5$]
- $\bullet^3$ strategy: substitution   ss
- $\bullet^4$ solve equation   pd
- $\bullet^5$ interpret result [state function]   ic

An extended response question that needs a calculator

⑫ (a) Show that a root to the equation $x^3 - x^2 - 3x - 5 = 0$ exists
   between $x = 2$ and $x = 3$.   *2 marks*
   (b) Find this root correct to 2 decimal places.   *3 marks*

$f(2) = -7 \Rightarrow f(2) < 0$
$f(3) = 4 \Rightarrow f(3) > 0$

So somewhere between $x = 2$ and $x = 3$, $f(x) = 0$.
So a root exists in the range $2 < x < 3$ ... and is closer to $x = 3$.
$f(2 \cdot 7) = -0 \cdot 707 ... \Rightarrow$ root is in the range $2 \cdot 7 < x < 3$
$f(2 \cdot 8) = 0 \cdot 712 ... \Rightarrow$ root is in the range $2 \cdot 7 < x < 2 \cdot 8$ ... and about halfway ...
$f(2 \cdot 75) = -0 \cdot 0156 ... \Rightarrow$ root is in the range $2 \cdot 75 < x < 2 \cdot 8$
$f(2 \cdot 76) = 0 \cdot 1269 ... \Rightarrow$ root is in the range $2 \cdot 75 < x < 2 \cdot 76$
... and closer to $2 \cdot 75$
$f(2 \cdot 751) = -0 \cdot 0014 ... \Rightarrow$ root is in the range $2 \cdot 751 < x < 2 \cdot 76$
$f(2 \cdot 752) = 0 \cdot 0127 ... \Rightarrow$ root is in the range $2 \cdot 751 < x < 2 \cdot 752$

To 2 decimal places $2 \cdot 751 = 2 \cdot 752 = 2 \cdot 75$.

The root of the equation $x = 2 \cdot 75$ correct to 2 decimal places.

Marking scheme

(a)

● ^{1} appreciate the requirements for proof   *ss*

● ^{2} evaluate polynomial twice and compare to zero   *pd*

(b)

● ^{3} determine approx roots of polynomial   *ss*

● ^{4} process data   *pd*

● ^{5} justify and express solution to required degree   *ic*

Topic Tips

Communication marks are often lost in questions in this field.

When asked for the nature of roots for example, be explicit about the conditions you are seeking. Don't write $b^2 - 4ac = 0$ when you really want $b^2 - 4ac > 0$. Many candidates automatically write $b^2 - 4ac = 0$ whether it is wanted or not.

When performing a synthetic division, draw to the examiner's attention the significance of a remainder of zero (this is a separate communication mark) ...

State the linear and the polynomial factor you find when the remainder is zero (again, a separate communication mark).

9 Basic Integration

What you should know

You are expected to know the following facts.

- The meaning of:
    - (i) integral ... see below;
    - (ii) integrate ... the process of finding an integral;
    - (iii) constant of integration ... see below;
    - (iv) definite integral... see below;
    - (v) limits of integration... see below;
    - (vi) indefinite integral ... see below;
    - (vii) area under a curve ... is the area trapped between the curve and the x-axis and the ordinates defined by the limits a and b.

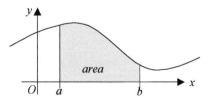

- If $f(x) = F'(x)$ then $\int f(x)\ dx = F(x) + C$ where C is the constant of integration ... is called an indefinite integral. No limits to the integral are given.

Related problems

You should be able to use this knowledge to solve the following kinds of problems.

- Integrate functions of the form $f(x) = ax^n$ for all rational n, except $n = -1$ by applying the rule $\int ax^n \ dx = \dfrac{ax^{n+1}}{n+1} + c$ where c is a constant.

- Integrate the sum or difference of such functions.

Example

$$\int x^4 + x + 1 + \sqrt{x} + \frac{1}{x^3} dx$$

   Make each term of the form ax^n

$$\int x^4 + x + 1 + x^{\frac{1}{2}} + x^{-3} \ dx$$

   Now apply the rule...

$$\frac{x^5}{5} + \frac{x^2}{2} + \frac{x^1}{1} + \frac{x^{\frac{3}{2}}}{\frac{3}{2}} + \frac{x^{-2}}{-2} + c$$

   ... and tidy up

$$\frac{x^5}{5} + \frac{x^2}{2} + x + \frac{2x^{\frac{3}{2}}}{3} - \frac{1}{2x^2} + c$$

Always remember to include c, the constant of integration.

- You may need to simplify an expression to prepare it for integration

Example

Integrate $\dfrac{2x^3 - x^2 + 1}{\sqrt{x}}$

Split it into three separate fractions ... and express the surd with a fractional index.

$$\frac{2x^3}{x^{\frac{1}{2}}} - \frac{x^2}{x^{\frac{1}{2}}} + \frac{1}{x^{\frac{1}{2}}}$$

Apply the laws of indices ... $a^m \div a^n = a^{m-n}$

$$2x^{\frac{5}{2}} - x^{\frac{3}{2}} + x^{-\frac{1}{2}}$$

... and apply the rule

$$\frac{2x^{\frac{7}{2}}}{\frac{7}{2}} - \frac{x^{\frac{5}{2}}}{\frac{5}{2}} + \frac{x^{\frac{1}{2}}}{\frac{1}{2}} + c$$

... and tidy up

$$\frac{4x^{\frac{7}{2}}}{7} - \frac{2x^{\frac{5}{2}}}{5} + 2x^{\frac{1}{2}} + c$$

- You will need to evaluate definite integrals.

  Example

  Evaluate $\displaystyle\int_0^1 3x^2 + \sqrt{x} + 1 \ dx$

  Prepare for integration,

  $\displaystyle\int_0^1 3x^2 + x^{\frac{1}{2}} + 1 \ dx$

  integrate, mindful of the limits...

  $\left[\dfrac{3x^3}{3} + \dfrac{x^{\frac{3}{2}}}{\frac{3}{2}} + x\right]_0^1$

  ... and tidy up.

  $\left[x^3 + \dfrac{2x^{\frac{3}{2}}}{3} + x\right]_0^1$

  Apply the rule ... if $f(x) = F'(x)$ then $\displaystyle\int_a^b f(x) \ dx = F(b) - F(a)$

  $\left[1^3 + \dfrac{2.1^{\frac{3}{2}}}{3} + 1\right] - \left[0^3 + \dfrac{2.0^{\frac{3}{2}}}{3} + 0\right]$

  ... and evaluate

  $1 + \frac{2}{3} + 1 - 0 = 2\frac{2}{3}$.

- Calculate the area bounded by the curve $y = f(x)$, the lines $x = a$, $x = b$ and the x-axis; $a < b$.

  (i)   where the curve does not cross the x-axis in the interval $a \le x \le b$

  we evaluate $\displaystyle\int_a^b f(x) \ dx$

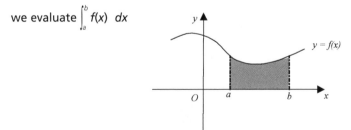

  A negative answer indicates the area lies below the x-axis. If you get a negative answer, don't just ignore it. To gain the interpretation mark you should explain it. Suppose the evaluation gives -12 then you respond with 'the area is 12 units^2 and is below the x-axis'.

> **!** Always give a justification for a negative answer when you evaluate areas ... and why you may be subsequently 'ignoring' the sign.

(ii) where the curve crosses the x-axis in the interval $a \leq x \leq b$ at a point $(c, 0)$

we evaluate $\int_a^c f(x)\ dx$ and $\int_c^b f(x)\ dx$ separately.

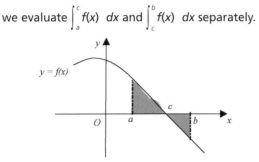

One of the values will be negative ... which you will explain as being below the x-axis ... before adding the two magnitudes, e.g. one value is 4 and the other is –6. You respond by writing 'The area is made up of two parts one which is 4 units^2 and above the x-axis, one 6 units^2 and below the x-axis ... a total of 10 units^2'.

> If you had been simply asked to evaluate the definite integral then no split is required; negatives may 'cancel out' positives; answers can be negative.

- Determine the area bounded by two curves.
  If the two curves, $y = f(x)$ and $y = g(x)$ intersect at $x = a$ and $x = b$ with no other intersections between these two points and $f(x)$ is the 'upper' function

  in the interval $a < x < b$ then the area $= \int_a^b (f(x) - g(x))\ dx$.

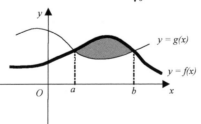

Where the two curves intersect more than once in the interval under consideration then each smaller area has to be considered separately. You can continue to consider the same integral as above, with different limits but the signs will switch as $f(x)$ changes from being the 'upper' to the 'lower' function. Mention this rather than lose a communication mark.

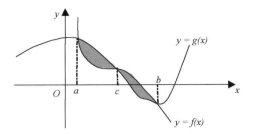

Note that being above or below the x-axis doesn't matter here.

- Solve differential equations of the form $\frac{dy}{dx} = f(x)$ to find an expression for

  y in terms of x. To evaluate the constant of integration, you will have to be given at least one pair of corresponding values of x and y.

  Example

  Find $f(x)$ such that $f'(x) = x^2 + 3x + 1$ and $f(0) = 1$.

  $$f(x) = \int x^2 + 3x + 1 \ dx = \frac{x^3}{3} + \frac{3x^2}{2} + x + c$$

  $$f(0) = \frac{0^3}{3} + \frac{3.0^2}{2} + 0 + c = 1 \Rightarrow c = 1$$

  So the required function is $f(x) = \frac{x^3}{3} + \frac{3x^2}{2} + x + 1$.

Objective questions

$$\int ax^n \ dx = \frac{ax^{n+1}}{n+1} + c$$

❶ Integrate $x^3 + \frac{1}{2x^2}$

A $\frac{x^4}{2} - \frac{1}{2x^3} + c$

B $\frac{x^4}{4} - \frac{1}{2x} + c$

C $\frac{x^2}{2} - \frac{1}{6x^3} + c$

D $\frac{x^2}{2} - \frac{1}{6x^3} + c$

Rough working

$$\int x^3 + \frac{1}{2}x^{-2}dx = \frac{x^4}{4} + \frac{1}{2} \cdot \frac{x^{-1}}{-1} + c = \frac{x^4}{4} - \frac{1}{2x^1} + c$$

Choose option B

Note that integrations can be checked by differentiating. You should get the original expression back.

❷ Find $\int \dfrac{\sqrt{x}}{x}\ dx$

A $-\dfrac{2x^{-3/2}}{3} + c$

B $-\dfrac{2}{3\sqrt{x^3}} + c$

C $-\dfrac{2}{\sqrt{x}} + c$

D $2\sqrt{x} + c$

Rough working

$\int \dfrac{\sqrt{x}}{x}\ dx = \int x^{-1/2}\ dx = \dfrac{x^{1/2}}{1/2} + c = 2\sqrt{x} + c$

Choose option D

Again differentiation can provide a quick answer check.

❸ Curves $y = f(x)$ and $y = g(x)$ intersect at (a, b) and (c, d) as shown. Which expression will give the shaded area?

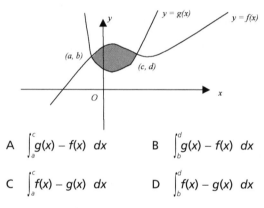

A $\displaystyle\int_a^c g(x) - f(x)\ dx$

B $\displaystyle\int_b^d g(x) - f(x)\ dx$

C $\displaystyle\int_a^c f(x) - g(x)\ dx$

D $\displaystyle\int_b^d f(x) - g(x)\ dx$

Rough working

Area $= \displaystyle\int_{lower\ x}^{upper\ x}$ *upper function – lower function dx*

Eliminate B and D ... they both use *y*-values for limits.
Eliminate A ... it treats $g(x)$ as the upper function.

Choose option C

❹ The sketch shows the function $y = x^3$.

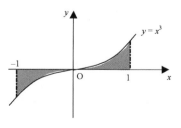

The shaded area can be calculated from which of these expressions?

(1) $\int_{-1}^{1} x^3 \, dx$ (2) $\int_{-1}^{0} x^3 \, dx + \int_{0}^{1} x^3 \, dx$ (3) $2\int_{0}^{1} x^3 \, dx$

A (1) only   B (2) only
C (3) only   D (2) and (3)

Rough working

$$\int_{-1}^{1} x^3 \, dx = \left[\frac{x^4}{4}\right]_{-1}^{1} = \frac{1}{4} - \frac{1}{4} = 0$$

$$\int_{-1}^{0} x^3 \, dx + \int_{0}^{1} x^3 \, dx = \left[\frac{x^4}{4}\right]_{-1}^{0} + \left[\frac{x^4}{4}\right]_{0}^{1} = 0 - \frac{1}{4} + \frac{1}{4} \, 0 = 0$$

There is an area to be seen ... so both of these can be eliminated.
As the examiners do not ask 'trick' questions then '(3) only' must be the answer.

Choose option C

As a check, given time, do the third integral.

This makes sense ... $y = x^3$ is symmetrical.

$$2\int_{0}^{1} x^3 \, dx = 2\left[\frac{x^4}{4}\right]_{0}^{1} = 2\left[\frac{1}{4}\right] = \frac{1}{2}$$

Extended response questions that don't need a calculator

❺ Calculate the area trapped between the curves $y = x^2 + x - 4$ and $y = 3x - 1$.   *5 marks*

(*Response*) ──

First we establish where the two curves intersect.

$y = f(x)$ intersects $y = g(x)$ where $f(x) = g(x)$

$x^2 + x - 4 = 3x - 1$
$\Rightarrow x^2 - 2x - 3 = 0$
$\Rightarrow (x + 1)(x - 3) = 0$
$\Rightarrow x = -1$ or 3

A sketch helps you establish which is the 'upper' function ...

Area = $\int_{-1}^{3}$ upper $-$ lower dx

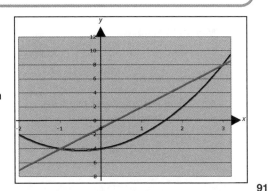

You have to make a statement like this in your response to show your selection of strategy. Without the statement it may be impossible for the examiner to decide which strategy you took ... resulting in the loss of a mark or two.

$$\text{Area} = \int_{-1}^{3} 3x - 1 - (x^2 + x + 4) \; dx = \int_{-1}^{3} 3 + 2x - x^2 \; dx$$

$$= \left[3x + x^2 - \frac{x^3}{3} \right]_{-1}^{3} = (9 + 9 - 9) - (-3 + 1 + \frac{1}{3})$$

$$= 10\tfrac{2}{3} \text{ square units}$$

Marking scheme

- •¹   determine points of intersection of line and curve   *ss*
  [identify points]
- •²   determine area between two curves   *ss*
  [identify form of integral that has to be done]
- •³   integrate   *pd*
- •⁴   use limits   *pd*
- •⁵   complete calcs. and communicate answer   *ic*

6   A function, f, is such that $f'(x) = \dfrac{1}{\sqrt{x}}$.

The curve $y = f(x)$ passes through the point $(4, 6)$.

Find the equation of this curve.   *4 marks*

(**Response**)

Let the examiner know that integration is your strategy.

$$y = \int \frac{1}{\sqrt{x}} \; dx = \int x^{-\frac{1}{2}} \; dx$$

$$= 2x^{\frac{1}{2}} + c$$

Interpret the given fact explicitly: when $x = 4$, $y = 6$.

$$y = 2\sqrt{x} + c$$
$$\Rightarrow 6 = 2 \times \sqrt{4} + c$$
$$\Rightarrow c = 2$$

Marshal the facts.

The required equation is $y = 2\sqrt{x} + 2$.

Marking scheme

- •¹   solve equations of the form   *ss*
- •²   integrate   *pd*
- •³   use the given point to find the constant   *ss*
- •⁴   bring together the findings to give answer   *ic*

These strategies will be further explored with other functions in unit 3.

Topic Tips

Notation and convention should not be ignored in this topic. It is fairly common for candidates to omit the dx when writing the integral. The constant of integration should never be forgotten. When solving simple differential equations the constant plays an active part. Missing it would mean missing the majority of the marks. When finding the area between two curves always tell the examiner your intentions, e.g.

by writing Area = $\int$ upper function − lower function dx.

Even written in this form you secure the strategy mark.

10 Trigonometric Formulae

What you should know

You are expected to know the following facts.

- Radian measure

> $180° = \pi$ radians or $1° = \frac{\pi}{180}$ radians

- How to solve trigonometric equations in a given interval.

  Example, level C
  $2\cos 2x = 1, 0 \leq x \leq \pi$.

  Example, level A/B
  $\cos^2 2x = 1; 3\sin^2 x + 7\sin x - 6 = 0, 0 \leq x \leq 2\pi$.

- Know how to use the addition formulae and double angle formulae where the following are given:

  $\sin(A \pm B) = \sin A \cos B \pm \cos A \sin B$
  $\cos(A \pm B) = \cos A \cos B \mp \sin A \sin B$
  $\sin 2A \qquad = 2\sin A \cos A$
  $\cos 2A \qquad = \cos^2 A - \sin^2 A$
  $\cos 2A \qquad = 2\cos^2 A - 1$
  $\cos 2A \qquad = 1 - 2\sin^2 A$

In the above identities A and B can be replaced by anything:

(i) replacing B by A produces the double angle formulae;

(ii) replacing B by 2A leads to triple angle formulae ... etc.;

(iii) replacing A by ½ in the double angle formulae leads to useful further identities which have been classified as A/B grade, e.g.

$$\sin x = 2 \sin \frac{x}{2} \cos \frac{x}{2} \qquad \cos x = \cos^2 \frac{x}{2} - \sin^2 \frac{x}{2}$$

$$\cos x = 2 \cos^2 \frac{x}{2} - 1 \qquad \cos x = 1 - 2\sin^2 \frac{x}{2}$$

Example, level C $\sin(P + 2Q) = \sin P \cos 2Q + \cos P \sin 2Q$

Example, level A/B $\sin x = 2\sin \frac{x}{2} \cos \frac{x}{2}$

Related problems

You should be able to use this knowledge to solve the following kinds of problems.

- Apply trigonometric formulae in the solution of geometric problems.

  (i) solution of triangles

  In this context, you have to know the sine rule, the cosine rule and other trigonometric facts learned at Credit level.

$$\frac{a}{\sin A} = \frac{b}{\sin B} = \frac{c}{\sin C} \qquad \cos A = \frac{b^2 + c^2 - a^2}{2bc}$$

$$\sin^2 A + \cos^2 A = 1 \qquad \tan A = \frac{\sin A}{\cos A}$$

Exact values of the trigonometric ratios derived from the square and the equilateral triangle may be used.

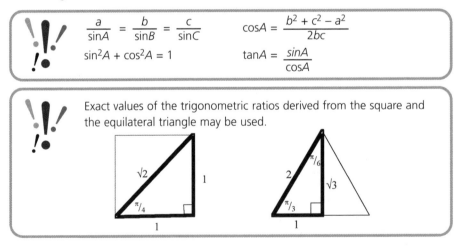

For example, it would not be unreasonable to be asked to prove that $a = \cos \theta + \sqrt{3} \sin \theta$ given the following triangle.

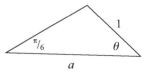

It brings together all these points in one question:

The third angle $= \pi - (\% + \theta)$ ...   working in radians

$$\frac{a}{\sin(\pi - (\% + \theta))} = \frac{1}{\sin\%}$$ ...   using the sine rule

$$\Rightarrow \frac{a}{\sin(\% + \theta)} = \frac{1}{\frac{1}{2}} = 2$$ ...   using $\sin(\pi - A) = \sin A$

$$\Rightarrow a = 2\sin(\% + \theta)$$

$$\Rightarrow a = 2\sin\%\cos\theta + 2\cos\%\sin\theta$$ ...   using compound angle formula

$$\Rightarrow a = 2.\tfrac{1}{2}.\cos\theta + 2.\tfrac{\sqrt{3}}{2}\sin\theta$$ ...   using exact values

$$\Rightarrow a = \cos\theta + \sqrt{3}\sin\theta$$ ...   Q.E.D.

Remember that the sine rule and cosine rule are NOT given in the Higher formulae sheet.

(ii) **three-dimensional situations**

In these problems the most common approach would be to look for right-angled triangles within which the rules of trigonometry can be applied.

For example:

A cuboid is constructed so that the angle that the space diagonal AD makes with the base, θ, is equal to the angle the face diagonal AC makes with the edge AB (see diagram).

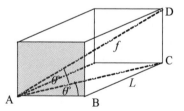

Prove that $L = \tfrac{1}{2}f\sin(2\theta)^\circ$ where L is the length of the cuboid and f is the length of the space diagonal.

From the diagram we can identify triangle ABC as right-angled and that
$\sin\theta = \frac{L}{AC} \Rightarrow AC = \frac{L}{\sin\theta}$
We can also identify triangle ACD as right-angled and that
$\cos\theta = \frac{AC}{f} \Rightarrow AC = f\cos\theta$
AC was chosen as a common side to the two triangles.
Equating the expressions for AC ... $f\cos\theta = \frac{L}{\sin\theta}$
$L = f\sin\theta\cos\theta$
$\Rightarrow L = \tfrac{1}{2}f.2.\sin\theta\cos\theta$ ... getting the form $2\sin A\cos A$
$\Rightarrow L = \tfrac{1}{2}f\sin 2\theta$ ... to use the double angle formula

(iii) 2D and 3D problems which include the need to use compound angle formulae. The principle behind most of these problems is that the sum, or difference of two angles is relatively easy to generate where the sines and cosines of the two angles, in their context, are easy to get, e.g.

$x = A + B$　　　　　　　$x = A - B$

In the context of the problem below you're expected to have Standard Grade knowledge.

The diagram shows a tangent kite of a circle whose radius is 5 cm. The tangents AB and AD are 12 cm long and intersect at an angle of $x°$.

The radii intersect at an angle of $y°$.

Calculate the exact value of
(i) $\sin x°$ (ii) $\cos y°$.

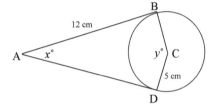

The initial pieces of knowledge used are:
(a) AC is an axis of symmetry (kite),
(b) $\angle ABC = \angle ADC = 90°$ (tangent/radius).
From this we can compute that
AC = 13 cm (Pythagoras' Theorem)
The diagram looks like this.

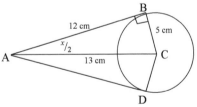

From the diagram we can get the values $\sin(\frac{x}{2})° = \frac{5}{13}$ and $\cos(\frac{x}{2})° = \frac{12}{13}$.

Now using the 'double' angle formula $\sin 2A = 2\sin A\cos A$ with $2A = x$
we get $\sin x° = 2\sin(\frac{x}{2})°\cos(\frac{x}{2})°$　　... this is a level A/B interpretation.
$\Rightarrow \sin x° = 2 \times \frac{5}{13} \times \frac{12}{13}$
$\Rightarrow \sin x° = \frac{120}{169}$

(iv) Finding $\cos y°$ can be found in a similar fashion using
$\cos y° = \cos^2(\frac{x}{2})° - \sin^2(\frac{x}{2})°$. ... this is a level A/B interpretation.
$\quad = (\frac{5}{13})^2 - (\frac{12}{13})^2$
$\quad = \frac{25}{169} - \frac{144}{169}$
$\quad = -\frac{119}{169}$

- Solve trigonometric equations using addition formulae.

  For example:
  (a)   show that $\sin(x + 60) = \frac{1}{2} \sin x° + \frac{\sqrt{3}}{2}\cos x°$,
  (b)   hence solve the equation $\sin x° + \sqrt{3} \cos x° = 1$, $0 \leq x \leq 360$.

  (a) $\sin(x + 60) = \sin x° \cos 60° + \cos x° \sin 60°$ ... using the compound angle
        formula $\sin(x + 60) = \frac{1}{2} \sin x° + \frac{\sqrt{3}}{2}\cos x°$ ... using the exact values (see
                                                                         above)

  (b) $\sin x° + \sqrt{3} \cos x° = 1$
        $\Rightarrow \frac{1}{2} \sin x° + \frac{\sqrt{3}}{2}\cos x° = \frac{1}{2}$
        $\Rightarrow \sin(x + 60) = \frac{1}{2}$ ... using the result from above
        $\Rightarrow x + 60 = 30, 180 - 30 = 150, 360 + 30 = 390, 360 + 150 = 510, ...$
        $\Rightarrow x = -30, 90, 330, 450 ...$
        Since $0 \leq x \leq 360$ then the solutions are 90° and 330°.

It is at the point where the inverse sine is taken that the list of solutions is generated. Given the equation $\sin(f(x))° = a$, there will be an answer from your calculator $f(x) = \sin^{-1}a$; there will be a solution generated by the symmetry of the sine wave, $f(x) = 180 - \sin^{-1}a$ and there will be all the solutions generated by adding 360° onto any solution you've found ... $f(x) = 360n + \sin^{-1}a$ and $f(x) = 360n + (180 - \sin^{-1}a)$ where n is any integer. It is a common mistake to leave the generation of multiple answers until x is known.

Solving	First Solution [calculator]	Second Solution [symmetry]	Periodicity 1 [adding 360s to calc]	Periodicity 2 [adding 360s to symm.]
$\sin x = a$	$x = \sin^{-1}a$	$x = 180 - \sin^{-1}a$	$x = 360n + \sin^{-1}a$	$x = 360n + 180 - \sin^{-1}a$
$\cos x = a$	$x = \cos^{-1}a$	$x = 360 - \cos^{-1}a$	$x = 360n + \cos^{-1}a$	$x = 360n + 360 - \cos^{-1}a$
$\tan x = a$	$x = \tan^{-1}a$	$x = 180 + \tan^{-1}a$	$x = 360n + \tan^{-1}a$	$x = 360n + 180 + \tan^{-1}a$

The standard way most people remember these facts is to picture this diagram.

Sin $180 - x$	All
Tan $180 + x$	Cos $360 - x$

The diagram lets you know where the trig ratios are positive and also what the symmetry for each particular ratio is.

- Solve trigonometric equations using double angle formulae.

  There are two standard types of equation to solve in this context:

  (a) Using sin 2A ... which leads to factorising by means of a common factor.

  For example: solve $\sin 2x - \sqrt{3} \cos x = 0$, $0 \le x \le 2\pi$.

  The first step is to cope with the double angle:
  $2 \sin x \cos x - \sqrt{3} \cos x = 0$
  $\Rightarrow \cos x(2 \sin x - \sqrt{3}) = 0$
  $\Rightarrow \cos x = 0$ or $(2 \sin x - \sqrt{3}) = 0$
  $\cos x = 0 \Rightarrow x = \frac{\pi}{2}, 2\pi - \frac{\pi}{2}, \dots$ and other values outside interval
  $\sin x = \frac{\sqrt{3}}{2} \Rightarrow x = \frac{\pi}{3}, \pi - \frac{\pi}{3} \dots$ and other values outside interval
  So list of solutions in the given interval are $\frac{\pi}{3}, \frac{\pi}{2}, \frac{2\pi}{3}, \frac{3\pi}{2}$.

  (b) Using cos 2A ... which leads to solving a quadratic equation in either sin x or cos x.

  For example: solve $\cos 2x° - 2 \cos x° = 2$, $0 \le x \le 360$.

  The equation has a term in $\cos 2x°$ and another term involving the single angle. You chose your formula for coping with the double angle according to the latter term.

  If it is $\sin x°$ then choose $\cos 2x° = 1 - 2 \sin^2 x°$.
  If it is $\cos x°$ then choose $\cos 2x° = 2 \cos 2x° - 1$.
  Thus we have $(2 \cos^2 x° - 1) - 2 \cos x° = 2$
              $\Rightarrow 2 \cos^2 x° - 2 \cos x° - 3 = 0$.
  Using the quadratic formula and solving for $\cos x°$

  $$\cos x = \frac{-(-2) \pm \sqrt{(-2)^2 - 4.2.(-3)}}{2.2} = \frac{2 \pm \sqrt{28}}{4}$$

  $\cos x = 1 \cdot 82288$ or $-0 \cdot 82288$.
  The first value has no solutions because $-1 \le \cos x° \le 1$
  so $x° = \cos^{-1}(-0 \cdot 82288) = 145 \cdot 4°$ or $214 \cdot 6°$.

  As in the example above, it's good practice to be explicit about why you have rejected certain avenues.
  In assessments, solutions would be on a given interval, e.g. $0 \le \theta \le 2\pi$.
  It would be possible for the following A/B question to appear.
  Solve $\sin x - \sqrt{3} \cos \left(\frac{x}{2}\right) = 0$, $0 \le x \le 2\pi$.

       $2 \sin \left(\frac{x}{2}\right) \cos \left(\frac{x}{2}\right) - \sqrt{3} \cos \left(\frac{x}{2}\right) = 0$
       $\Rightarrow \cos \left(\frac{x}{2}\right) \left(2 \sin \left(\frac{x}{2}\right) - \sqrt{3}\right) = 0$
       $\Rightarrow \cos \left(\frac{x}{2}\right) = 0$ or $\left(2 \sin \left(\frac{x}{2}\right) - \sqrt{3}\right) = 0$
       $\cos \left(\frac{x}{2}\right) = 0 \Rightarrow \frac{x}{2} = \frac{\pi}{2}, 2\pi - \frac{\pi}{2}, \dots$ and other values outside interval
       $\Rightarrow x = \pi \dots$ and other values outside interval
       $\sin \left(\frac{x}{2}\right) = \frac{\sqrt{3}}{2} \Rightarrow \frac{x}{2} = \frac{\pi}{3}, \pi - \frac{\pi}{3} \dots$ and other values outside interval
       $\Rightarrow x = \frac{2\pi}{3}, \frac{4\pi}{3} \dots$ and other values outside interval
       So list of solutions in the given interval are $\frac{2\pi}{3}, \pi, \frac{4\pi}{3}$.

Objective questions

❶ Given that the $\sin x° = \frac{3}{5}$ and that $0 < x < 90$, what is the exact value of $\sin 2x°$?

A $\frac{6}{5}$

B $\frac{9}{10}$

C $\frac{12}{25}$

D $\frac{24}{25}$

Rough working

Since x is acute, we may put it in a right-angled triangle.

From this we see the third side is 4 ... and that $\cos x = \frac{4}{5}$

$\sin 2x° = 2 \sin x° \cos x° = 2 \times \frac{3}{5} \times \frac{4}{5} = \frac{24}{25}$.

Choose option D

❷ What is $\cos(\frac{\pi}{2} + x)$ equal to?

A $\cos x$

B $-\cos x$

C $\sin x$

D $-\sin x$

Rough working

Expanding we get $\cos(\frac{\pi}{2} + x) = \cos \frac{\pi}{2} \cos x - \sin \frac{\pi}{2} \sin x = 0.\cos x - 1.\sin x$
$= -\sin x$.

Choose option D

❸ $15 = 45 - 30$. What is the **exact** value of $\sin 15°$.

A $\dfrac{1+\sqrt{3}}{2\sqrt{2}}$ B $\dfrac{1-\sqrt{3}}{2\sqrt{2}}$

C $\dfrac{\sqrt{3}-1}{2\sqrt{2}}$ D $\dfrac{\sqrt{3}+1}{2\sqrt{2}}$

Rough working

$\sin 15° = \sin(45 - 30)° = \sin 45° \cos 30° - \cos 45° \sin 30°$

$= \frac{1}{\sqrt{2}} \times \frac{\sqrt{3}}{2} - \frac{1}{\sqrt{2}} \times \frac{1}{2} = \dfrac{\sqrt{3}-1}{2\sqrt{2}}$

For future reference: $15° = 45° - 30°$; $75° = 45° + 30°$;
$120° = 180° - 60$, $105° = 90° + 15°$ etc.

Choose option C

④ Given that cos $a° = \frac{3}{5}$ and that $0 < x < 90$,
simplify $\sin(x + a)° + \sin(x - a)°$.

A $\sin x°$
B $\frac{3}{5} \sin x°$
C $\frac{4}{5} \sin x°$
D $\frac{6}{5} \sin x°$

Rough working

$\sin(x + a)° + \sin(x - a)°$
   $= \sin x° \cos a° + \cos x° \sin a° + \sin x° \cos a° - \cos x° \sin a°$
   $= 2 \sin x° \cos a°$
   $= 2 \times \frac{3}{5} \times \sin x°$
   $= \frac{6}{5} \sin x°$

Choose option D

⑤ In the diagram sin $a° = \frac{3}{5}$ and cos $a° = \frac{4}{5}$.
What is the value of sin $b°$?

A $\frac{33}{65}$
B $\frac{63}{65}$
C $\frac{16}{65}$
D $\frac{56}{65}$

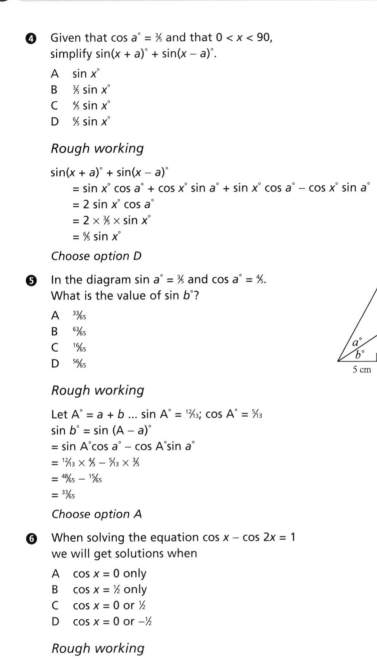

12 cm

$a°$
$b°$

5 cm

Rough working

Let $A° = a + b$ ... sin $A° = \frac{12}{13}$; cos $A° = \frac{5}{13}$
$\sin b° = \sin (A - a)°$
$= \sin A° \cos a° - \cos A° \sin a°$
$= \frac{12}{13} \times \frac{4}{5} - \frac{5}{13} \times \frac{3}{5}$
$= \frac{48}{65} - \frac{15}{65}$
$= \frac{33}{65}$

Choose option A

⑥ When solving the equation cos $x -$ cos $2x = 1$
we will get solutions when

A   cos $x = 0$ only
B   cos $x = \frac{1}{2}$ only
C   cos $x = 0$ or $\frac{1}{2}$
D   cos $x = 0$ or $-\frac{1}{2}$

Rough working

$\cos x - \cos 2x = 1$
$\Rightarrow \cos x - 2\cos^2 x + 1 = 1$

$$\Rightarrow \cos x - 2\cos^2 x = 0$$
$$\Rightarrow \cos x(1 - 2\cos x) = 0$$
$$\Rightarrow \cos x = 0 \text{ or } \cos x = \tfrac{1}{2}$$

Choose option C

❼ In the diagram, the sine of $a°$ = ⅓.
What is the tangent of $b°$?

A $\dfrac{1}{2\sqrt{2}}$ 　　　　B $-\dfrac{1}{2\sqrt{2}}$

C $\dfrac{1}{2}$ 　　　　D $-\dfrac{1}{\sqrt{2}}$

Rough working

$\tan b° = \tan(360 - a°) = -\tan a°$

$a°$ is acute and so there exists a triangle containing $a°$
where $\sin a° = \tfrac{1}{3}$.
By Pythagoras' Theorem the missing side is $\sqrt{8} = 2\sqrt{2}$.
From this we see that $\tan a° = \tfrac{1}{2}\sqrt{2}$
and so $\tan b° = -\tfrac{1}{2}\sqrt{2}$.

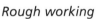

Choose option B

A note for extended response questions that don't need a calculator

In this particular topic, when a question appears in the non-calculator paper, it usually involves using **exact** values.

These are obtained from either the half-square or the half-equilateral triangle (see above) or the sine or cosine of some related angle(s) whose sine, cosine or tangent is given.

For example, in question 7 (above) you are given that $\sin a° = \tfrac{1}{3}$ and a diagram which lets you see that $a°$ is acute. This allows you to make a convenient triangle, find the missing side by Pythagoras' Theorem and establish the exact values of $\sin a°$, $\cos a°$ and $\tan a°$ as needed.

This is generally considered a level C skill.
However, you may be told that $a°$ is obtuse
(or reflex) and this would transform the
question to a level A/B skill.
Suppose $\sin a° = \tfrac{1}{3}$ and $a°$ was obtuse.
Then the strategy to use is to draw $a°$ in the
coordinate plane as shown here.

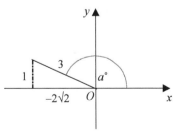

Now you can deduce that the x-coordinate is
$-2\sqrt{2}$ and read off the sine, cosine or tangent of $a°$.

Extended response questions that don't need a calculator

8 An alleyway is 12 units wide. Two
ladders are in the alleyway as
shown in the diagram, one 13
units long and the other 15 units
long.

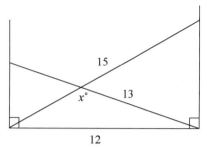

What is the exact value of sin x°, where x° is the angle at which
the two ladders cross?   *5 marks*

(**Response**)────────────────────────────────

For communication purposes, the
diagram should be redrawn into
your answer booklet.
Use Pythagoras' Theorem to
calculate the missing sides of the
two triangles.
Since $x°$ is not in a right-angled
triangle, you should label *related*
angles which are.

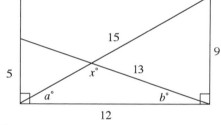

$\sin x° = \sin(180 - (a + b))° = \sin(a + b)°$
$= \sin a° \cos b° + \cos a° \sin b°$
$= \dfrac{9}{15} \cdot \dfrac{12}{13} + \dfrac{12}{15} \cdot \dfrac{5}{13}$
$= \dfrac{56}{65}$

─────────────────────────────────⟨   ⟩

Marking scheme

●¹ identify crucial aspect of problem   *ss*
     [find the missing sides]
●² interpret a collection of facts   *ic*
     [to relate x, a and b and find the missing sides]
●³ apply trig formulae to soln of geometric formula   *ss*
●⁴ apply knowledge of trig formulae   *pd*
●⁵ interpret diagram [to obtain values]   *ic*

9 Calculate the exact value of tan x in the diagram below. *4 marks*

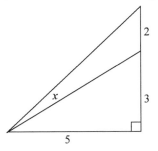

(**Response**)

The same note applies here. Since *x* is not in a right-angled triangle you have to relate it to angles which are. Redraw the diagram, labeling the required angles.

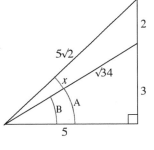

$\tan x = \tan (A - B)$

$$\frac{\sin(A - B)}{\cos(A - B)} = \frac{\sin A \cos B - \sin A \cos B}{\cos A \cos B + \sin A \sin B}$$

$$= \frac{\tfrac{5}{\sqrt{2}}.\tfrac{5}{\sqrt{34}} - \tfrac{5}{\sqrt{2}}.\tfrac{3}{\sqrt{34}}}{\tfrac{5}{\sqrt{2}}.\tfrac{5}{\sqrt{34}} + \tfrac{5}{\sqrt{2}}.\tfrac{3}{\sqrt{34}}}$$

$$= \frac{5 - 3}{5 + 3} = \frac{2}{8} = \frac{1}{4}$$

Marking scheme

- ●¹ identify crucial aspect of problem **ss**
  [find the missing sides]
- ●² apply trig formulae to solution of geometric formula **ss**
- ●³ apply knowledge of trig formulae **pd**
- ●⁴ interpret diagram [to obtain values] **ic**

Similar problems can show up wherever there is a well-known theorem relating angles.

The following problem, though difficult, brings many points together.

10 The circle has a diameter, AB, 25 cm long. The chord BD is 15 cm long and the chord AC is 24 cm.

Calculate the **exact** value of cos ∠CEB.

5 marks

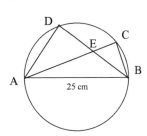

(Response) ────────────────────────────────────

First sketch the diagram into the answer book and mark 'givens' and things you can immediately deduce (such as right angles). Highlight what you want.

∠ADB = ∠ACB = 90° (angles in a semicircle)
The angle under question is not in a right-angled triangle with enough information... relate it to ones that are.

∠CEB = ∠EAB + ∠EBA (exterior angle = sum of interior opposites)
∠CEB = ∠CAB + ∠DBA (same angles ... arms extended)
By Pythagoras' Theorem AD = 20 and BC = 7
$$\cos \angle CEB = \cos(\angle CAB + \angle DBA)$$
$$= \cos \angle CAB \cos \angle DBA - \sin\angle CAB \sin\angle DBA$$
$$= \frac{24}{25}\cdot\frac{15}{25} - \frac{7}{25}\cdot\frac{20}{25} = \frac{360}{625} - \frac{140}{625} = \frac{220}{625} = \frac{44}{125}$$

Marking scheme

- •¹   identify crucial aspect of problem   *ss*
  [angles in a semicircle, exterior angle]
- •²   interpret a collection of facts   *ic*
  [to relate angles and find the missing sides]
- •³   apply trig formulae to solution of geometric formula   *ss*
- •⁴   apply knowledge of trig formulae   *pd*
- •⁵   interpret diagram [to obtain values]   *ic*

⑪   This is a sketch of two trigonometric functions.

$y = \sin 2x$ and $y = \sqrt{3}\, \cos x$, $0 \le x \le 2\pi$

Find where the curves intersect in this interval.

4 marks

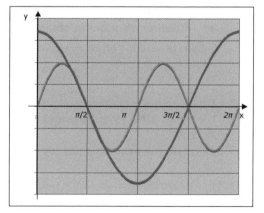

(Response)

At first sight it looks as though the curves intersect twice: once when $x = \frac{\pi}{6}$ and once where $x = \frac{3\pi}{2}$. However, looks aren't good enough – if you just state this you will not earn any marks. At the front of the exam paper it reads, 'Answers obtained by readings from scale drawings will not receive any credit' and the diagram counts as a scale drawing.

The curves intersect when $\sin 2x = \sqrt{3} \cos x$

$\Rightarrow \sin 2x - \sqrt{3} \cos x = 0$

$\Rightarrow 2\sin x \cos x - \sqrt{3} \cos x = 0$

$\Rightarrow \cos x(2\sin x - \sqrt{3}) = 0$

$\Rightarrow \cos x = 0$ or $2\sin x - \sqrt{3} = 0$

$\Rightarrow \cos x = 0$ or $\sin x = \frac{\sqrt{3}}{2}$

$\Rightarrow x = \frac{\pi}{2}$ or $\frac{3\pi}{2}$ or $x = \frac{\pi}{3}$ or $\pi - \frac{\pi}{3} = \frac{2\pi}{3}$

Marking scheme

- •^{1} use the double angle formula   *ss*
- •^{2} factorise   *pd*
- •^{3} obtain first angles from each equation   *pd*
- •^{4} obtain the rest of the angles in the range   *pd*

Watch out for the following.

(i)   The appearance of $x = \frac{\pi}{6}$ and $x = \frac{3\pi}{2}$ without supportive working will not receive any credit.

(ii) If more than four solutions are given then mark 4 will not be awarded.

(iii) If you don't give your answer in radians you'll lose a mark.

⑫ Relative to a suitable set of axes, two orbiting crafts trace out paths that can be modeled by $y = 4\cos 2x°$ and $y = 2\cos x° - 1$ where y is the height in suitable units and x is the distance round the orbit measured in degrees.

As can be seen in the diagram their paths cross at four places in one complete orbit. Calculate the values of x at which the paths cross.

5 marks

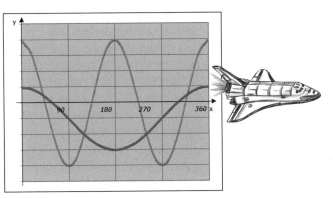

Response

The craft cross where $4\cos 2x° = 2\cos x° - 1$

$\Rightarrow 4\cos 2x° - 2\cos x° + 1 = 0$

$\Rightarrow 4(2\cos^2 x° - 1) - 2\cos x° + 1 = 0$

$\Rightarrow 8\cos^2 x° - 2\cos x° - 3 = 0$

$\Rightarrow (2\cos x° + 1)(4\cos x° - 3) = 0$

$\Rightarrow 2\cos x° + 1 = 0$ or $4\cos x° - 3 = 0$

$\Rightarrow \cos x° = -\frac{1}{2}$ or $\cos x° = \frac{3}{4}$

$\Rightarrow x° = 120°$ or $240°$ or $x° = 41·4°$ or $318·6°$ (correct to 1 d.p.) ——————⊂⊃

Marking scheme

- $\bullet^1$ use the double angle formula   *ss*
- $\bullet^2$ form quadratic in cos x *pd*
- $\bullet^3$ know to solve quadratic in cos x *ss*
- $\bullet^4$ obtain first angles from each equation   *pd*
- $\bullet^5$ obtain the rest of the angles in the range   *pd*

Note

If errors or omissions are made, marks 4 and 5 can be obtained by other means: mark 4 can be obtained for '$\cos x° = -\frac{1}{2} \Rightarrow x° = 120°$ or $240°$' and mark 5 for '$\cos x° = \frac{3}{4} \Rightarrow x° = 41·4°$ or $318·6°$ (correct to 1 d.p.)'. Alternatively, mark 4 can be obtained for '$\cos x° = -\frac{1}{2}$ and $\cos x° = \frac{3}{4}$' and mark 5 for '$x° = 120°$ or $240°$ or $x° = 41·4°$ or $318·6°$ (correct to 1 d.p.)' so you should be very clear about laying out your solutions to this type of question.

Topic Tips

Practice working with radians. Some of the questions – if they are done in degrees – are simply a Credit skill. Radians were invented to make life easier. If you work in degrees and then try to switch to radians you are adding another layer to the problem ... another source of error. If you forget to switch back YOU WILL lose marks.

Credit skills can still be called on. Remember the sine rule and the cosine rule. Remember that trig formulae are given at the front of the exam paper. Get familiar with them. Transcription errors are penalised at Higher. The formulae use A and B as the variables. Remember that these can be exchanged for anything including ½ (which would indeed make a problem A/B grade). If you let B = 2A you would find a formula for sin 3A etc.

When solving trig equations, the multiple answers appear at the first instance of applying the inverse trig function ... NOT at the end of the question.

When given a geometric diagram and asked for exact values of trig ratios look for angle connections like A + B, A – B, 360 – (A + B) etc.

What you should know

Related problems

Objective questions

Extended response questions that don't need a calculator

What you should know

You are expected to know the following facts.

- The equation of the circle centre (a, b) and radius r is $(x - a)^2 + (y - b)^2 = r^2$.
- The equation $x^2 + y^2 + 2gx + 2fy + c = 0$ represents a circle centre $(-g, -f)$ and radius $\sqrt{(g^2 + f^2 - c)}$ provided $g^2 + f^2 - c > 0$.

Both of these facts will be given to you in the examination in the formula sheet.

Related problems

You should be able to use this information to solve the following kinds of problems.

- Determining the equation of a circle.
  (i) The simplest case is when the centre and radius are given.
     It's just a simple case of substitution.

  Example
  Find the equation of a circle with centre (4, −2) and radius 5.

  (Response)

  Equation of circle: $(x - 4)^2 + (y - (-2))^2 = 5^2$
  $$\Rightarrow (x - 4)^2 + (y + 2)^2 = 25.$$

(ii) If the centre and one point on the circumference are given:
Use the two points to find the radius ... making use of the distance formula $r = \sqrt{(x_2 - x_1)^2 + (y_2 - y_1)^2}$.

Thereafter it's like case (i).

Example
Find the equation of a circle with centre (4, −2) and passing through the point (7, 2).

Response

Centre = (4, −2)

radius = $r = \sqrt{(4 - 7)^2 + (-2 - 2)^2} = \sqrt{9 + 16} = 5$

Equation of circle: $(x - 4)^2 + (y - (-2))^2 = 5^2$

                   $\Rightarrow (x - 4)^2 + (y + 2)^2 = 25$

(iii) If two points are given that are diametrically opposite:
establish the centre as the midpoint between the given points.

$$\left(\frac{x_2 + x_1}{2}, \frac{y_2 + y_1}{2}\right) = (4, -2)$$

Use the centre and one of the points to calculate the radius.
Thereafter it's like case (i).

Example
Find the equation of a circle with diameter AB where A is (1, −6) and B is (7, 2).

Response

Centre = $\left(\dfrac{1 + 7}{2}, \dfrac{-6 + 2}{2}\right) = (4, -2)$

Radius = $r = \sqrt{(4 - 7)^2 + (-2 - 2)^2} = \sqrt{9 + 16} = 5$

Equation of circle: $(x - 4)^2 + (y - (-2))^2 = 5^2$

                   $\Rightarrow (x - 4)^2 + (y + 2)^2 = 25$

(iv) If three points are given.
Select any two points. They will define a chord of the required circle.
Find the equation of the perpendicular bisector of this chord.

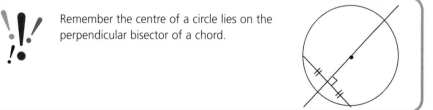

Remember the centre of a circle lies on the perpendicular bisector of a chord.

Pair off a different two points. They will define another chord.
Find the equation of the perpendicular bisector of this chord.
Find where these two perpendicular bisectors intersect – this is the
centre of the circle.
Use the centre and any one point to find the radius. Thereafter it's like case (i).

Example
Find the equation of a circle which passes through the three points A(1, –6),
B(0, 1) and C(4, 3).

(**Response**)

$m_{AB} = \dfrac{1 - (-6)}{0 - 1} = -7 \Rightarrow m_\perp = \dfrac{1}{7}$; midpoint AB ($\frac{1}{2}$, $-\frac{5}{2}$)

Equation of perpendicular bisector: $y + \frac{5}{2} = \frac{1}{7}(x - \frac{1}{2})$

$m_{BC} = \dfrac{3 - 1}{4 - 0} = \dfrac{2}{4} \Rightarrow m_\perp = -2$; midpoint BC (2, 2)

Equation of perpendicular bisector: $y - 2 = -2(x - 2)$

      Bisectors intersect when $\frac{1}{7}(x - \frac{1}{2}) - \frac{5}{2} = -2(x - 2) + 2$

      $\Rightarrow 2(x - \frac{1}{2}) - 35 = -28(x - 2) + 28$

      $\Rightarrow 2x - 1 - 35 = -28x + 56 + 28$

      $\Rightarrow 30x = 120$

      $\Rightarrow x = 4$

      $\Rightarrow y = -2(4 - 2) + 2 = -2$

Centre of circle is (4, –2)

radius $= r = \sqrt{(4 - 0)^2 + (-2 - 1)^2} = \sqrt{16 + 9} = 5$ ... using point B.

Equation of Circle: $(x - 4)^2 + (y - (-2))^2 = 5^2$

                 $\Rightarrow (x - 4)^2 + (y + 2)^2 = 25.$

● Determining the points at which a given line intersects a given circle.

Example
Where does the line $y = 3x - 9$ intersect the circle $x^2 + y^2 - 8x + 4y - 5 = 0$?

(**Response**)

Substitute the linear expression for y into the equation of the circle:

$x^2 + (3x - 9)^2 - 8x + 4(3x - 9) - 5 = 0$

$\Rightarrow x^2 + 9x^2 - 54x + 81 - 8x + 12x - 36 - 5 = 0$

$\Rightarrow 10x^2 - 50x + 40 = 0$

$\Rightarrow x^2 - 5x + 4 = 0$

$\Rightarrow (x - 4)(x - 1) = 0$

$\Rightarrow x = 4$ or 1

$\Rightarrow y = 3$ or -6

So line cuts the circle at (4, 3) and (1, –6).

- Determining whether a given line is a tangent to a given circle.
  Level C: when looking for points of intersection discover there is only one and deduce tangency.
  Level A/B: example given by the examiners (the SQA) ... The line with equation $x - 3y = k$ is a tangent to the circle
  $x^2 + y^2 - 6x + 8y + 15 = 0$. Find the two possible values of k.

When the linear expression is substituted into the equation of the circle, the resultant equation is a quadratic. By considering the discriminant we can decide on how the line interacts with the circle.

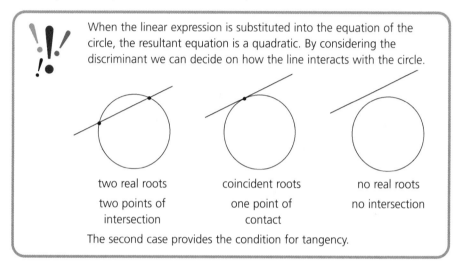

two real roots	coincident roots	no real roots
two points of intersection	one point of contact	no intersection

The second case provides the condition for tangency.

Response

It is easier to substitute for x in this case ... $x = k + 3y$
$(k + 3y)^2 + y^2 - 6(k + 3y) + 8y + 15 = 0$
$\Rightarrow k^2 + 6ky + 9y^2 + y^2 - 6k - 18y + 8y + 15 = 0$
$\Rightarrow 10y^2 + (6k - 10)y + (k^2 - 6k + 15) = 0$
For the line to be a tangent, the discriminant of this equation must equal zero.
$a = 10, b = (6k - 10), c = (k^2 - 6k + 15)$
$b^2 - 4ac = 0$
$\Rightarrow (6k - 10)^2 - 4 \times 10 \times (k^2 - 6k + 15) = 0$
$\Rightarrow 36k^2 - 120k + 100 - 40k^2 + 240k - 600 = 0$
$\Rightarrow -4k^2 + 120k - 500 = 0$
$\Rightarrow k^2 - 30k + 125 = 0$
$\Rightarrow (k - 5)(k - 25) = 0$
$\Rightarrow k = 5$ or 25

Note: there will be marks available for knowing to use the discriminant and for setting its value to zero. These marks are usually not given if you don't state the evidence obviously, i.e. we should see the line '$b^2 - 4ac = 0$ for coincident roots' to secure the marks.

- Determining whether two circles touch each other.
  Two circles will *touch* if the distance between their centres, d, is equal to the sum of their radii, or the difference between their radii.

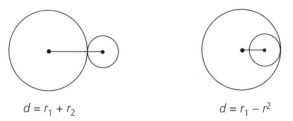

$$d = r_1 + r_2 \qquad\qquad\qquad d = r_1 - r^2$$

Two circles will intersect at two points when $r_1 - r_2 < d < r_1 + r_2$.

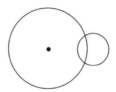

The centre of one circle will lie on the other circle when $d = r_1$ or $d = r_2$.

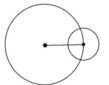

Two circles are concentric when $d = 0$.

- Determining whether a given equation represents a circle.
  At its simplest level, this will entail examining $g^2 + f^2 - c$.
  If this is less than or equal to zero then the equation does not represent a circle.

Why? … because $r = \sqrt{(g^2 + f^2 - c)}$ and you can't have a circle with a radius which is zero or non-real.

However, this could be a good time for the examiner to test your skills at completing the square.

Example

(a) By completing the square twice, express $x^2 + y^2 + 2x + 6y + 15$ in the form $(x - a)^2 + (y - b)^2 = c$.

(b) Does $x^2 + y^2 + 2x + 6y + 15 = 0$ represent a circle? Explain your decision, making reference to part (a).

Remember, $x^2 = -1$ has no real solution.

(Response) ————————————————————————————

(a) $x^2 + y^2 + 2x + 6y + 15$
     $= x^2 + 2x + y^2 + 6y + 15$
     $= (x + 1)^2 - 1^2 + (y + 3)^2 - 3^2 + 15$
     $= (x + 1)^2 + (y + 3)^2 + 5$

(b) $x^2 + y^2 + 2x + 6y + 15 = 0$
     $\Rightarrow (x + 1)^2 + (y + 3)^2 + 5 = 0$
     $\Rightarrow (x + 1)^2 + (y + 3)^2 = -5$

This would represent a circle if $r^2 = -5$.
But no real number fits this description.
So this can't be a circle. ———————————————————————

If no evidence of the steps of completing the square are apparent, the marks for part (a) are not available.
No marks will be available a candidate who does not use his/her findings from part (a) to establish the findings of part (b).

Objective questions

❶   Which of the following statements are true about $x^2 + y^2 + 4x - 6y + 9 = 0$?

     A   It represents a circle with centre $(2, -3)$ and radius 4.
     B   It represents a circle with centre $(-2, 3)$ and radius 4.
     C   It represents a circle with centre $(-2, 3)$ and radius 2.
     D   It does not represent a circle.

Rough working

$g = 2$, $f = -3$ and $c = 9$
centre $(-g, -f)$ is $(-2, 3)$
radius $= \sqrt{(g^2 + f^2 - c)} = \sqrt{(4 + 9 - 9)} = 2$

Choose option C

❷ Which points lie inside the circle $(x - 1)^2 + (y - 2)^2 = 4$?

A $(0, 0)$ only
B $(0, 0)$ and $(2, 3)$
C $(2, 3)$ only
D $(2, 1)$ and $(2, 3)$

All points on the circle satisfy the equation $(x - 1)^2 + (y - 2)^2 = 4$.
All points **outside** the circle satisfy the inequation
$(x - 1)^2 + (y - 2)^2 > 4$.
All points **inside** the circle satisfy the inequation
$(x - 1)^2 + (y - 2)^2 < 4$.

Rough working

$(0, 0)$: $(0 - 1)^2 + (0 - 2)^2 > 4$ ... outside circle
$(2, 1)$: $(2 - 1)^2 + (1 - 2)^2 = 2 < 4$ ... inside circle
$(2, 3)$: $(2 - 1)^2 + (3 - 2)^2 = 2 < 4$ ... inside circle

Choose option D

❸ $(x - 2)^2 + (y - 3)^2 = 4$ and $x^2 + y^2 - 4x - 6y + 17 = 0$
What is true about this pair of equations?

A They represent the same circle.
B They represent concentric circles.
C They both represent circles of radius 2.
D They are not both circles.

Rough working

The first equation represents a circle of radius 2 and centre $(2, 3)$.
The second has $g = -2$, $f = -3$ and $c = 17$: $g^2 + f^2 - c = 4 + 9 - 17 < 0$.
So this does not represent a circle.

Choose option D

❹ For what value(s) of k is $y = k$ a tangent to the circle
$x^2 + y^2 - 10x + 2y + 17 = 0$?

A   4 and −2
B   −4 and 2
C   2 and 8
D   −2 and −8

Rough working

The centre is (5, −1). The radius is $\sqrt{(25 + 1 - 17)} = 3$.
$y = k$ is a line parallel to the x-axis ... 3 units above or below the centre
i.e. $y = -1 + 3 = 2$ or $y = -1 - 3 = -4$

Alternatively, substituting into the original gives
$x^2 + k^2 - 10x + 2k + 17 = 0$
$\Rightarrow x^2 - 10x + k^2 + 2k + 17 = 0$
$b^2 - 4ac = 0$ for tangency.
$\Rightarrow 100 - 4(k^2 + 2k + 17) = 0$
$\Rightarrow k^2 + 2k - 8 = 0$
$\Rightarrow (k - 2)(k + 4) = 0$
$\Rightarrow k = 2$ or -4

Choose option B

❺ What is the equation of a circle with centre (3, −4) passing through the origin?

A $(x - 3)^2 + (y + 4)^2 = 25$
B $(x + 3)^2 + (y - 4)^2 = 25$
C $(x - 3)^2 + (y + 4)^2 = 16$
D $(x + 3)^2 + (y - 4)^2 = 16$

Rough working

The distance from the centre (3, −4) to the origin $= \sqrt{((3 - 0)^2 + (4 - 0)^2)} = 5$.
This, of course, is the radius. Thus the equation of the circle is
$(x - 3)^2 + (y + 4)^2 = 25$.

Choose option A

❻ What is the equation of a circle with centre (−6, 7) that has the y-axis as a tangent?

A $(x - 6)^2 + (y + 7)^2 = 36$
B $(x + 6)^2 + (y - 7)^2 = 36$
C $(x - 6)^2 + (y + 7)^2 = 49$
D $(x + 6)^2 + (y - 7)^2 = 49$

Rough working

A quick sketch helps ...
... we see that the radius must be 6 units.
So the equation is $(x + 6)^2 + (y - 7)^2 = 6^2$

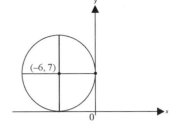

Choose option B

Extended response questions that don't need a calculator

❼ A circle has the equation $x^2 + y^2 - 2x + 14y - 119 = 0$.
The point A(6, 5) lies on this circle.
Find the equation of the tangent to the circle at this point.   *4 marks*

（Response）————————————————————————

Centre of the circle is C(1, −7)

$m_{AC} = \dfrac{-7 - 5}{1 - 6} = \dfrac{12}{5}$

$m_{tangent} = -\dfrac{5}{12}$

$y - 5 = -\dfrac{5}{12}(x - 6)$ ——————————————

Marking scheme

- •^{1} interpret the equation of a circle   *ic*
        [find the centre]
- •^{2} know how to find gradient [of radius]   *ss*
- •^{3} know how to find gradient of perpendicular   *ss*
- •^{4} recall and apply $y - b = m(x - a)$ *ic*

❽ The centre of a circle is C(6, −2). The circle passes through the point
(10, −5).
(a) What is the equation of the circle?   *3 marks*
(b) The line with equation $x - 7y + 5 = 0$ cuts through this circle
at B and D.
B is closer to the origin. Find the coordinates of B and D.   *4 marks*
(c) Find the equation of the tangent to the circle at B.   *3 marks*

（Response）————————————————————————

(a) The radius = $\sqrt{[(10 - 6)^2 + (-5 - (-2))^2]} = 5$.
     Thus the equation is $(x - 6)^2 + (y + 2)^2 = 25$.

(b) $x - 7y + 5 = 0 \Rightarrow x = 7y - 5$

Substituting this into the equation of the circle

gives $(7y - 5 - 6)^2 + (y + 2)^2 = 25$

$\Rightarrow (7y - 11)^2 + (y + 2)^2 = 25$

$\Rightarrow 49y^2 - 154y + 121 + y^2 + 4y + 4 - 25 = 0$

$\Rightarrow 50y^2 - 150y + 100 = 0$

$\Rightarrow y^2 - 3y + 2 = 0$

$\Rightarrow (y - 2)(y - 1) = 0$

$\Rightarrow y = 1$ or 2

$\Rightarrow x = 2$ or 9

Points of intersection are B(2, 1) and D(9, 2).

(c) Centre of the circle is C(6, −2)

$$m_{BC} = \frac{-2 - 1}{6 - 2} = \frac{-3}{4}$$

$$m_{tangent} = \frac{4}{3}$$

$$y - 1 = \frac{4}{3}(x - 2)$$

Marking scheme

- •^{1} interpret statements   *ss*
  [appreciating significance of given data]
- •^{2} know to find radius   *ss*
- •^{3} state equation of circle   *ic*
- •^{4} determine pts. of intersection of a line with a circle   *ss*
- •^{5} substitute and form quadratic   *pd*
- •^{6} determine roots of equation   *ss*
- •^{7} express as coordinates and identify B and D   *ic*
- •^{8} know how to find gradient   *ss*
  [of radius]
- •^{9} know how to find gradient of perpendicular   *ss*
- •^{10} recall and apply $y - b = m(x - a)$ *ic*

9 The vertices of a triangle have coordinates P(−9, −3), Q(8, 14) and R(16, 2).

(a) Calculate the equation of the perpendicular bisector of PQ.   *4 marks*

(b) Calculate the equation of the circle that passes through the three vertices of the triangle.   *6 marks*

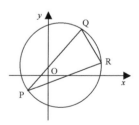

Note: This circle is called the *circumcircle* of PQR and could be referred to as this in the exam. The centre of a circumcircle lies at the intersection of the perpendicular bisectors of the sides.

(a) The mid point of PQ is $\left(\dfrac{-9 + 8}{2}, \dfrac{-3 + 14}{2}\right) = \left(-\dfrac{1}{2}, \dfrac{11}{2}\right)$

$m_{PQ} = \dfrac{14 + 3}{8 + 9} = 1 \Rightarrow m_{\perp} = -1$

Equation of perpendicular bisector of PQ: $y - \dfrac{11}{2} = -\left(x + \dfrac{1}{2}\right)$

i.e. $y = -x + 5$

(b) The mid point of QR is $\left(\dfrac{16 + 8}{2}, \dfrac{2 + 14}{2}\right) = (12, 8)$

$m_{QR} = \dfrac{14 - 2}{8 - 16} = \dfrac{12}{-8} = -\dfrac{3}{2} \Rightarrow m_{\perp} = \dfrac{2}{3}$

Equation of perpendicular bisector of QR: $y - 8 = \dfrac{2}{3}(x - 12)$

i.e. $y = \dfrac{2}{3}x$

These lines will intersect where $-x + 5 = \dfrac{2}{3}x \Rightarrow x = 3$

$\Rightarrow y = 2$

Centre of the required circle is C(3, 2).

Radius $= CP = \sqrt{((3 + 9)^2 + (2 + 3)^2)} = \sqrt{169} = 13$.

Equation of circle: $(x - 3)^2 + (y - 2)^2 = 169$.

Marking scheme

- $\bullet^1$ find the mid-point   ss
  [appreciating significance of given data]
- $\bullet^2$ find gradient of chord   ss
- $\bullet^3$ find gradient of perpendicular   ss
- $\bullet^4$ recall and apply $y - b = m(x - a)$ ic
- $\bullet^5$ repeat strategy for another chord   ss
- $\bullet^6$ process the data   pd
- $\bullet^7$ state the equation   ic
- $\bullet^8$ marshal the facts   ss
- $\bullet^9$ solve simultaneously to find intersection   ss
- $\bullet^{10}$ state the equation of the circle   ic

⑩ Two circles are drawn with equations $x^2 + y^2 + 6x + 10y + 9 = 0$ and $x^2 + y^2 - 4x - 14y - 11 = 0$.

(a) Express both these equations in the form
     $(x - a)^2 + (y - b)^2 = r^2$.   3 marks
(b) Prove these two circles have a single point of contact.   3 marks
(c) Calculate the gradient of the tangent passing through this
     point of contact.   3 marks

(Response) ──

(a) Circle 1: Centre $(-3, -5)$; radius $= \sqrt{(9 + 25 - 9)} = 5$

Equation $(x + 3)^2 + (y + 5)^2 = 25$

Circle 2: Centre $(2, 7)$; radius $= \sqrt{(4 + 49 + 11)} = 8$

Equation $(x - 2)^2 + (y - 7)^2 = 64$

(b) The distance between the centres $= \sqrt{((2 + 3)^2 + (7 + 5)^2)} = 13$.

Note that the sum of the two radii $= 5 + 8 = 13$.

The centres are exactly the sum of the radii apart.

So the circles are touching.

(c) The line joining the centres has a gradient $\dfrac{7 + 5}{2 + 3} = \dfrac{12}{5}$.

The tangent will be perpendicular to this and so have a gradient $\dfrac{-5}{12}$.

──────────────────────────────────── ⬭

Marking scheme

- •¹ interpret equation of circle finding centre and radius *ss*
- •² express in required form *ss*
- •³ repeat for second circle *ic*
- •⁴ determine distance between centres *pd*
- •⁵ know to evaluate sum of radii *ss*
- •⁶ draw conclusion *ic*
- •⁷ find gradient of line joining centres *pd*
- •⁸ identify crucial aspects i.e. tangent perpendicular to radius *ss*
- •⁹ find gradient of a perpendicular, recalling $m_1 m_2 = -1$ *ic*

Topic Tips

Finding the centre and radius of a circle by geometric means using Standard Grade skills. This can often catch you out – especially in circle problems. Don't look for difficulties. The examiner is not interested in asking obscure or trick questions. Often you can make use of simple 2D vectors. To find a point diametrically opposite another, the simplest thing is to realise that the vector from the point to the centre will be the same as the vector from the centre to the desired point.

Keep in mind that the given formula $x^2 + y^2 + 2gx + 2fy + c = 0$ only represents a circle when certain conditions are met
... when $g^2 + f^2 - c > 0$.

12 Vectors in Three Dimensions

What you should know

Related problems

Objective questions

An extended response question that doesn't need a calculator

An extended response question that needs a calculator

An extended response question that doesn't need a calculator

An A/B short response question

What you should know

You are expected to know the following facts.

● Know the terms:

vector – a vector is a quantity which has magnitude and direction;

magnitude (length) – the magnitude of a vector u is the length of any directed line segment that represents u and is denoted by $|u|$;

scalar – any pure number is called a scalar to differentiate it from a vector;

scalar multiple – if k is a scalar then ku is called the scalar multiple of u; It is a *vector* parallel to u with a magnitude of $k|u|$

position vector – of a point P, usually denoted by p, is a vector which is represented by the directed line segment $\overrightarrow{OP}$ where O is the origin;

unit vector – a vector whose magnitude is 1;

directed line segment – a line is considered to extend to infinity in both directions. Any single part of this line is called a line segment. When we define a direction along this segment we call it a directed line segment, e.g. AB may be a line segment. If we go from A to B then we would refer to the directed line segment $\overrightarrow{AB}$. If we go from B to A then we get $\overrightarrow{BA}$;

component – relative to a set of coordinate axes (at Higher grade this is the x-direction, the y-direction and the z-direction), a vector can be described by the steps taken in each of the three directions.

In general a vector will look like $\begin{pmatrix} x \\ y \\ z \end{pmatrix}$.

Each of these numbers is a component of the vector.

As a convention, the components of the vector are denoted using subscripts. i.e.

$$a = \begin{pmatrix} a_1 \\ a_2 \\ a_3 \end{pmatrix}$$

Scalar product ... the scalar product of the vectors *a* and *b* is a scalar denoted by *a.b* (and is thus often referred to as the dot product). It can be calculated from the defining formula $a.b = |a|\,|b|\cos\theta$, where θ is the angle between the vectors when they are tail-to-tail.

● Know the properties of vector addition and multiplication ...

$$\begin{pmatrix} a_1 \\ a_2 \\ a_3 \end{pmatrix} + \begin{pmatrix} b_1 \\ b_2 \\ b_3 \end{pmatrix} = \begin{pmatrix} a_1 + b_1 \\ a_2 + b_2 \\ a_3 + b_3 \end{pmatrix}$$

$$\begin{pmatrix} 0 \\ 0 \\ 0 \end{pmatrix} \text{ is the zero vector}$$

$$\begin{pmatrix} a_1 \\ a_2 \\ a_3 \end{pmatrix} + \begin{pmatrix} -a_1 \\ -a_2 \\ -a_3 \end{pmatrix} = \begin{pmatrix} 0 \\ 0 \\ 0 \end{pmatrix} \text{ ... defining the negative of } \begin{pmatrix} a_1 \\ a_2 \\ a_3 \end{pmatrix}$$

$$k\begin{pmatrix} a_1 \\ a_2 \\ a_3 \end{pmatrix} = \begin{pmatrix} ka_1 \\ ka_2 \\ ka_3 \end{pmatrix} \text{ ... a scalar multiple}$$

$a.b = |a|\,|b|\cos\theta$ ... the scalar product.

● Know the properties of a vector by a scalar
$k(a + b) = ka + kb$.

● Know the formula for the distance between two points in three dimensional space, given $A(a_1, a_2, a_3)$ and $B(b_1, b_2, b_3)$ the distance
$AB = \sqrt{[(a_1 - b_1)^2 + (a_2 - b_2)^2 + (a_3 - b_3)^2]}$.

- Know the equality fact

$$\begin{pmatrix} a \\ b \\ c \end{pmatrix} = \begin{pmatrix} d \\ e \\ f \end{pmatrix} \Rightarrow a = d, \, b = e, \, c = f$$

- Know that if **u** and **v** are vectors that can be represented by parallel lines then **u** = k**v** where k is a constant.
- Know the converse is also true, i.e. if **u** = k**v** then **u** and **v** are parallel.
- Know that if A, P and B are collinear points such that
$$\frac{AP}{PB} = \frac{m}{n} \Rightarrow \overrightarrow{AP} = \frac{m}{n} \, \overrightarrow{PB}.$$
- Know the basis vectors $i = \begin{pmatrix} 1 \\ 0 \\ 0 \end{pmatrix}, j = \begin{pmatrix} 0 \\ 1 \\ 0 \end{pmatrix}, k = \begin{pmatrix} 0 \\ 0 \\ 1 \end{pmatrix}.$

- An understanding of both two- and three- dimensional vectors is expected. Any of the above rules which apply to 3D can be adapted to 2D by omitting the third component.
- Know the scalar product facts:
a.b = |a| |b| cos θ

$$\textbf{a.b} = a_1 b_1 + a_2 b_2 + a_3 b_3, \text{ where } a = \begin{pmatrix} a_1 \\ a_2 \\ a_3 \end{pmatrix} \text{ and } b = \begin{pmatrix} b_1 \\ b_2 \\ b_3 \end{pmatrix}$$

a.(b + c) = **a.b** + **a.c**
a.b.c has no meaning.

- If A is the point A(a_1, a_2, a_3) then its position vector $\overrightarrow{OA} = \begin{pmatrix} a_1 \\ a_2 \\ a_3 \end{pmatrix}.$

- If A and B are the points A(a_1, a_2, a_3) and B(b_1, b_2, b_3)
then vector $\overrightarrow{AB} = b - a = \begin{pmatrix} b_1 - a_1 \\ b_2 - a_2 \\ b_3 - a_3 \end{pmatrix}.$

Related problems

You should be able to use the above knowledge to solve the following kinds of problems.

- Interpreting a 3D coordinate system represented in 2D.
  The exam paper often contains a question where a 3D coordinate system is set up and you are asked to state the coordinates of a point in the picture.
  e.g. What are the coordinates of point A?

$\boxed{\text{Response}}$ ────────────────────────────────

A(4, 2, 3) ────────────────────────────────────

- Finding the magnitude of a vector.

  Given the vector $a = \begin{pmatrix} a_1 \\ a_2 \\ a_3 \end{pmatrix}$ then its magnitude $|a| = \sqrt{a_1^2 + a_2^2 + a_3^2}$.

- Add, subtract and find scalar multiples of vectors.

  Example

  Given $\begin{pmatrix} 3 \\ p \\ 4 \end{pmatrix} - 4\begin{pmatrix} 1 \\ 3 \\ q \end{pmatrix} = \begin{pmatrix} r \\ 15 \\ 16 \end{pmatrix}$, find p, q and r.

$\boxed{\text{Response}}$ ────────────────────────────────

$$\begin{pmatrix} 3 \\ p \\ 4 \end{pmatrix} - 4\begin{pmatrix} 1 \\ 3 \\ q \end{pmatrix} = \begin{pmatrix} r \\ 15 \\ 16 \end{pmatrix}$$

$$\Rightarrow \begin{pmatrix} 3 - 4 \\ p - 12 \\ 4 - 4q \end{pmatrix} = \begin{pmatrix} r \\ 15 \\ 16 \end{pmatrix}$$

$$\Rightarrow \begin{cases} 3 - 4 = r \\ p - 12 = 15 \\ 4 - 4q = 16 \end{cases}$$ which gives $r = -1$, $p = 27$ and $q = -3$. ──────

> When two vectors are equal, corresponding components are equal.

- Express a directed line segment in terms of the position vectors of its ends.

Example

Given that A is (1, 2, –4) and B is (–3, 5, –1)

then $\vec{AB} = \mathbf{b} - \mathbf{a} = \begin{pmatrix} -3 - 1 \\ 5 - 2 \\ -1 - (-4) \end{pmatrix} = \begin{pmatrix} -4 \\ 3 \\ 3 \end{pmatrix}$.

> This can be considered the result of a 'vector walk' ...

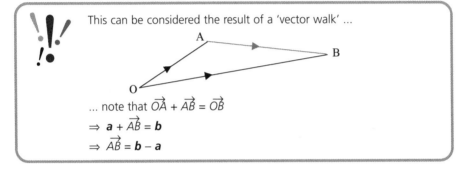

... note that $\vec{OA} + \vec{AB} = \vec{OB}$

$\Rightarrow \mathbf{a} + \vec{AB} = \mathbf{b}$

$\Rightarrow \vec{AB} = \mathbf{b} - \mathbf{a}$

- Work with directed line segments in 3D space to add or subtract vectors ('vector walks') and apply scalar product properties.

Example

ABCD is a regular tetrahedron of side 1 unit.

E divides $\vec{BD}$ in the ratio 2:1.

(a) Express $\vec{CE}$ in terms of the vectors $\mathbf{u}$, $\mathbf{v}$ and $\mathbf{w}$.

(b) Find the scalar product $\vec{CD}.\vec{CE}$.

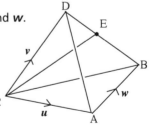

Response

(a) $\vec{CE} = \vec{CD} + \vec{DE}$

$= \vec{CD} + \frac{1}{3}\vec{DB}$

$= \vec{CD} - \frac{1}{3}\vec{BD}$

$= \vec{CD} - \frac{1}{3}(-\vec{AB} - \vec{CA} + \vec{CD})$

$$= \boldsymbol{v} - \tfrac{1}{3}(-\boldsymbol{w} - \boldsymbol{u} + \boldsymbol{v})$$
$$= \tfrac{1}{3}(2\boldsymbol{v} + \boldsymbol{w} + \boldsymbol{u})$$

(b) $\boldsymbol{v}.\tfrac{1}{3}(2\boldsymbol{v} + \boldsymbol{w} + \boldsymbol{u}) = \boldsymbol{v}.\tfrac{1}{3}(2\boldsymbol{v} + \overrightarrow{CB}) = \tfrac{1}{3}(2\boldsymbol{v}.\boldsymbol{v} + \boldsymbol{v}.\overrightarrow{CB})$
$$= \tfrac{1}{3}(2.1.1.\cos 0° + 1.1.\cos 60°)$$
$$= \tfrac{1}{3}(2 + \tfrac{1}{2})$$
$$= \tfrac{5}{6}$$

- Express and work with vectors in component form and use the basis vectors *i*, *j*, *k*.

Example

$$A = \begin{bmatrix} 2 \\ -1 \\ 6 \end{bmatrix} = \begin{bmatrix} 2 \\ 0 \\ 0 \end{bmatrix} + \begin{bmatrix} 0 \\ -1 \\ 0 \end{bmatrix} + \begin{bmatrix} 0 \\ 0 \\ 6 \end{bmatrix} = 2\begin{bmatrix} 1 \\ 0 \\ 0 \end{bmatrix} - 1\begin{bmatrix} 0 \\ 1 \\ 0 \end{bmatrix} + 6\begin{bmatrix} 0 \\ 0 \\ 1 \end{bmatrix} = 2\boldsymbol{i} - \boldsymbol{j} + 6\boldsymbol{k}$$

> The basis vectors are of unit magnitude and are mutually perpendicular to each other
> … so *i.j* = *j.k* = *k.i* = 0 and *i.i* = *j.j* = *k.k* = 1.

- Solve problems using the conditions for equality (equality fact already mentioned).

Example

There are four points in space.

A(0, −1, q), B(p, 3, $2q$), C(2, 1, $2p$) and D(q, 5, 5) where p and q are constants.
It is known that $\overrightarrow{AB} = \overrightarrow{CD}$.
Calculate the values of p and q.

Response

$$\overrightarrow{AB} = \boldsymbol{b} - \boldsymbol{a} = \begin{bmatrix} p \\ 3 \\ 2q \end{bmatrix} - \begin{bmatrix} 0 \\ -1 \\ q \end{bmatrix} = \begin{bmatrix} p \\ 4 \\ q \end{bmatrix}$$

$$\overrightarrow{CD} = \boldsymbol{d} - \boldsymbol{c} = \begin{bmatrix} q \\ 5 \\ 5 \end{bmatrix} - \begin{bmatrix} 2 \\ 1 \\ 2p \end{bmatrix} = \begin{bmatrix} q - 2 \\ 4 \\ 5 - 2p \end{bmatrix}$$

If these are equal then corresponding components are equal.

$\Rightarrow p = q - 2$ and $q = 5 - 2p$

$\Rightarrow p = 5 - 2p - 2$

$\Rightarrow 3p = 3$

$\Rightarrow p = 1$

$\Rightarrow q = 3$

- Prove that vectors are parallel.

> Given two vectors **a** and **b**, **a** is parallel to **b** if $\mathbf{a} = k\mathbf{b}$, $k \in \mathbb{R}$.

Example

A(3, −2, 2), B(5, −1, −1), C(−2, 0, 4) and D(4, 3, −5) are four points.
Prove that AB is parallel to CD.

(*Response*)

$$\vec{AB} = \mathbf{b} - \mathbf{a} = \begin{pmatrix} 5 \\ -1 \\ -1 \end{pmatrix} - \begin{pmatrix} 3 \\ -2 \\ 2 \end{pmatrix} = \begin{pmatrix} 2 \\ 1 \\ -3 \end{pmatrix}$$

$$\vec{CD} = \mathbf{d} - \mathbf{c} = \begin{pmatrix} 4 \\ 3 \\ -5 \end{pmatrix} - \begin{pmatrix} -2 \\ 0 \\ 4 \end{pmatrix} = \begin{pmatrix} 6 \\ 3 \\ -9 \end{pmatrix} = 3 \begin{pmatrix} 2 \\ 1 \\ -3 \end{pmatrix} = 3\vec{AB}$$

SO AB is parallel to CD.

You really have to draw it to the examiner's attention that $CD = 3\vec{AB}$ and follow through with a declared conclusion to pick up the marks.

- Determine whether three points with given coordinates are collinear.

    Note: In two dimensions, given the coordinates of three points A, B and C, you can show that AB and BC have the same gradient and hence 'run in the same direction'. Coupling that with the fact that B is a point common to both line segments is enough to prove A, B and C all lie on the same line. (Alternatively you could find the equation of the line defined by A and B and then prove C lies on it.)

    In three dimensions you have to show that AB and BC, say, pass the 'parallel' test and couple that with the fact that B is a common point to prove collinearity.

Example

Prove that A(1, 2, 4), B(7, 4, 0) and C(10, 5, −2) are collinear.

Response

$$\vec{AB} = b - a = \begin{bmatrix} 7 \\ 4 \\ 0 \end{bmatrix} - \begin{bmatrix} 1 \\ 2 \\ 4 \end{bmatrix} = \begin{bmatrix} 6 \\ 2 \\ -4 \end{bmatrix}$$

$$\vec{BC} = c - b = \begin{bmatrix} 10 \\ 5 \\ -2 \end{bmatrix} - \begin{bmatrix} 7 \\ 4 \\ 0 \end{bmatrix} = \begin{bmatrix} 3 \\ 1 \\ -2 \end{bmatrix} = \tfrac{1}{2}AB$$

So BC and AB are parallel.

But AB and BC have a point in common, namely B.

So AB and BC lie on the same straight line.

Therefore A, B and C are collinear.

A communication mark will be lost if you don't declare that the two line segments have a common point.

- Find scalar products.

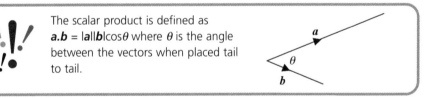

The scalar product is defined as
a.b = |**a**||**b**|cosθ where θ is the angle between the vectors when placed tail to tail.

It can also be shown that when $a = \begin{bmatrix} a_1 \\ a_2 \\ a_3 \end{bmatrix}$ and $b = \begin{bmatrix} b_1 \\ b_2 \\ b_3 \end{bmatrix}$ then

$a.b = a_1b_1 + a_2b_2 + a_3b_3$.

Proof

Using basis vectors: $a.b = (a_1i + a_2j + a_3k) . (b_1i + b_2j + b_3k)$

$= a_1b_1 i.i + a_1b_2 i.j + a_1b_3 i.k + a_2b_1 j.i + a_2b_2 j.j + a_2b_3 j.k + a_3b_1 k.i + a_3b_2 k.j + a_3b_3 k.k$

and using $i.j = j.k = k.i = 0$ and $i.i = j.j = k.k = 1$ we get

$a.b = a_1b_1 + a_2b_2 + a_3b_3$

As its name suggests, it is a *scalar*.

- Determine whether or not two vectors, in component form, are perpendicular.
  The strategy to do this depends on the fact that if $|a||b|\cos\theta = 0$ it can only be because $|a| = 0$, $|b| = 0$ or $\cos\theta = 0 \Rightarrow \theta = 90°$ (or 270°).
  If neither of the vectors are zero vectors then they must be mutually perpendicular.

  You may be asked to prove that one vector is perpendicular to another or you may be asked to find conditions that will ensure vectors are mutually perpendicular. In either case you should exhibit your knowledge of the theorem by stating it. You should make sure that you declare that the vectors are not zero vectors ... otherwise the implication that the vectors are perpendicular to each other would be invalid.

> If two non-zero vectors have a scalar product of zero then the vectors are mutually perpendicular.

Example 1
Prove that $u = 3i + 2j - 4k$ is perpendicular to $v = 2i + 3j + 3k$

(Response)
$u.v = u_1v_1 + u_2v_2 + u_3v_3 = 3.2 + 2.3 + (-4).3 = 6 + 6 - 12 = 0$
Since the scalar product is zero and neither of the vectors are zero vectors, the vectors must be mutually perpendicular.

Example 2
Two vectors are defined as $u = xi - j - 2k$ and $v = xi + xj + 3k$, $x > 0$
u and v mutually perpendicular. Find u and v.

(Response)
Note that neither u nor v is a zero vector.
To be mutually perpendicular, their scalar product must equal zero.
$u.v = u_1v_1 + u_2v_2 + u_3v_3 = x.x + (-1).x + (-2).3 = x^2 - x - 6$
$x^2 - x - 6 = 0$
$\Rightarrow (x - 3)(x + 2) = 0$
$\Rightarrow x = -2$ or 3.
We know $x > 0$ so $x = 3$
$\Rightarrow u = 3i - j - 2k$ and $v = 3i + 3j + 3k$

- Use scalar product to find the angle between two directed line segments.
  The strategy for determining this is most commonly used for finding the size of an angle in 3D space, say $\angle ABC$ when the coordinates of A, B and C are known.

Use is made of the definition of the scalar product $\mathbf{a.b} = |\mathbf{a}||\mathbf{b}|\cos\theta$.

Step 1: Use the coordinates to find the vectors $\overrightarrow{BA}$ and $\overrightarrow{BC}$ ... note the choice of vectors. We want an angle with its vertex at B. We select two vectors coming away from B. A common mistake is to select $\overrightarrow{AB}$ and $\overrightarrow{BC}$ and to calculate the supplement of the angle you want. You can avoid this error by making a point of sketching the situation ... always draw arrows emanating from the required vertex.

Step 2: Use the formula $\mathbf{a.b} = a_1b_1 + a_2b_2 + a_3b_3$ to get the value of the scalar product.

Step 3: Use the formula $|\mathbf{a}| = \sqrt{a_1^2 + a_2^2 + a_3^2}$ to find the magnitudes of both vectors.

Step 4: Use the formula $\cos\theta = \dfrac{\mathbf{a.b}}{|\mathbf{a}||\mathbf{b}|}$ to get θ

Example

Calculate the size of the angle $\angle PQR$ given the three points $P(1,2,4)$, $Q(-3, -4, 1)$ and $R(2, 1, -3)$.

Response

$$\overrightarrow{QP} = p - q = \begin{pmatrix} 1 \\ 2 \\ 4 \end{pmatrix} - \begin{pmatrix} -3 \\ -4 \\ 1 \end{pmatrix} = \begin{pmatrix} 4 \\ 6 \\ 3 \end{pmatrix}$$

$$\overrightarrow{QR} = r - q = \begin{pmatrix} 2 \\ 1 \\ -3 \end{pmatrix} - \begin{pmatrix} -3 \\ -4 \\ 1 \end{pmatrix} = \begin{pmatrix} 5 \\ 5 \\ -4 \end{pmatrix}$$

$$\overrightarrow{QP}.\overrightarrow{QR} = \begin{pmatrix} 4 \\ 6 \\ 3 \end{pmatrix} . \begin{pmatrix} 5 \\ 5 \\ -4 \end{pmatrix} = 20 + 30 + (-12) = 38$$

$$|\overrightarrow{QP}| = \sqrt{4^2 + 6^2 + 3^2} = \sqrt{61}$$

$$|\overrightarrow{QR}| = \sqrt{5^2 + 5^2 + (-4)^2} = \sqrt{66}$$

$$\cos\angle PQR = \frac{QP.QR}{|QP||QR|} = \frac{38}{\sqrt{61}\sqrt{66}} = 0\cdot599 \text{ (to 3 s.f.)}$$

$$\Rightarrow \angle PQR = 53\cdot2°$$

● Find a point which divides a line in a given ratio.
There are two cases of this problem.
(i) Where the point divides the line internally [level C]

In this diagram the point P divides the line AB internally in the ratio m:n.

We can write:

$$\frac{\overrightarrow{AP}}{\overrightarrow{PB}} = \frac{m}{n}$$

$$\Rightarrow n\overrightarrow{AP} = m\overrightarrow{PB}$$

$$\Rightarrow n(p - a) = m(b - p)$$

$$\Rightarrow np - na = mb - mp$$

$$\Rightarrow p(m + n) = na + mb$$

$$\Rightarrow p = \frac{na + mb}{m + n}$$

This is called the section formula and can be used to find P.
However, depending on your memory to come up with formulae can be a dangerous habit. It is usually safer to go through the process each time.

Example
Find the point P which divides the line AB internally in the ratio 2:3 given that A and B are the points (1, 2, 6) and (11, −3, 11) respectively.

(**Response**) ───────────────────────────────────

$$\frac{\overrightarrow{AP}}{\overrightarrow{PB}} = \frac{2}{3}$$

$$\Rightarrow 3\overrightarrow{AP} = 2\overrightarrow{PB}$$

$$\Rightarrow 3(p - a) = 2(b - p)$$

$$\Rightarrow 3p - 3a = 2b - 2p$$

$$\Rightarrow 5p = 3a + 2b$$

$$\Rightarrow p = \frac{1}{5}\left[3\begin{pmatrix} 1 \\ 2 \\ 6 \end{pmatrix} + 2\begin{pmatrix} 11 \\ -3 \\ 11 \end{pmatrix} \right] = \frac{1}{5}\begin{pmatrix} 25 \\ 0 \\ 40 \end{pmatrix}$$

$$\Rightarrow p = \begin{pmatrix} 5 \\ 0 \\ 8 \end{pmatrix}$$

P is the point (5, 0, 8). ───────────────────────────

Some people prefer to work with basis vectors. In this case the last three lines would become:

$\Rightarrow 5\boldsymbol{p} = 3(\boldsymbol{i} + 2\boldsymbol{j} + 6\boldsymbol{k}) + 2(11\boldsymbol{i} - 3\boldsymbol{j} + 11\boldsymbol{k})$

$\Rightarrow 5\boldsymbol{p} = 3\boldsymbol{i} + 6\boldsymbol{j} + 18\boldsymbol{k} + 22\boldsymbol{i} - 6\boldsymbol{j} + 22\boldsymbol{k}$

$\Rightarrow 5\boldsymbol{p} = 25\boldsymbol{i} + 0\boldsymbol{j} + 40\boldsymbol{k}$

$\Rightarrow \boldsymbol{p} = 5\boldsymbol{i} + 0\boldsymbol{j} + 8\boldsymbol{k}$

P is the point (5, 0, 8).

(ii) Where the point divides the line externally [level A/B].

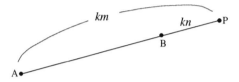

In this diagram the point P divides the line AB externally in the ratio *m:n*.

Note that the directed line segment $\overrightarrow{AP}$ runs in the opposite direction to $\overrightarrow{PB}$ ... and so should have opposite signs to denote this. Thus when we say P divides the line AB externally in the ratio *m:n*, we can state that

$$\frac{\overrightarrow{AP}}{\overrightarrow{PB}} = \frac{-m}{n}.$$

Apart from this negative sign, the rest of the argument is the same ... and it's level A/B.

Example

Find the point P which divides the line AB **externally** in the ratio 2:3 given that A and B are the points (1, 2, 6) and (11, −3, 11) respectively.

Response

$\dfrac{\overrightarrow{AP}}{\overrightarrow{PB}} = \dfrac{-2}{3}$

$\Rightarrow 3\overrightarrow{AP} = -2\overrightarrow{PB}$

$\Rightarrow 3(\boldsymbol{p} - \boldsymbol{a}) = -2(\boldsymbol{b} - \boldsymbol{p})$

$\Rightarrow 3\boldsymbol{p} - 3\boldsymbol{a} = -2\boldsymbol{b} + 2\boldsymbol{p}$

$\Rightarrow \boldsymbol{p} = 3\boldsymbol{a} - 2\boldsymbol{b}$

$\Rightarrow \boldsymbol{p} = 3\begin{bmatrix} 1 \\ 2 \\ 6 \end{bmatrix} - 2\begin{bmatrix} 11 \\ -3 \\ 11 \end{bmatrix}$

$\Rightarrow \boldsymbol{p} = \begin{bmatrix} -19 \\ 12 \\ -4 \end{bmatrix}$ P is the point (−19, 12, −4).

Note:
Although

$$\frac{-2}{3} = \frac{2}{-3} = -\frac{2}{3}$$

you're generally safer associating the negative sign with the numerator ... there's less chance of subsequent error.

The SQA give three specific notes about this topic.
(i) The section formula may be used to find the position vector of P but it is not required. As already suggested, more marks are lost in a badly-remembered formula than in simply arguing the point out each time.
(ii) 2D should be known as a special case of the 3D situation.
So any problem in 3D can be reduced to a 2D formula by considering the z-component to be zero.
In two dimensions:
- the magnitude of a vector would be given by $|a| = \sqrt{a_1^2 + a_2^2}$;
- the scalar product $\mathbf{a}.\mathbf{b} = a_1 b_1 + a_2 b_2$.

Example
Triangle ABC has vertices A(−3, 10), B(−2, 3) and C(1, 2).
Find the size of angle ∠ABC.

(Response)

Although there are many ways of getting this answer (e.g. using the distance formula and trigonometry), we could use the scalar product.

$$\overrightarrow{BA} = a - b = \begin{bmatrix} -3 \\ 10 \end{bmatrix} - \begin{bmatrix} -2 \\ 3 \end{bmatrix} = \begin{bmatrix} -1 \\ 7 \end{bmatrix}$$

$$\overrightarrow{BC} = c - b = \begin{bmatrix} 1 \\ 2 \end{bmatrix} - \begin{bmatrix} -2 \\ 3 \end{bmatrix} = \begin{bmatrix} 3 \\ -1 \end{bmatrix}$$

$$\overrightarrow{BA}.\overrightarrow{BC} = \begin{bmatrix} -1 \\ 7 \end{bmatrix}.\begin{bmatrix} 3 \\ -1 \end{bmatrix} = -3 - 7 = -10$$

$$|\overrightarrow{BA}| = \sqrt{(-1)^2 + 7^2} = \sqrt{50}$$

$$|\overrightarrow{BC}| = \sqrt{(-1)^2 + 3^2} = \sqrt{10}$$

$$\cos\angle ABC = \frac{-10}{\sqrt{500}}$$

$$\angle ABC = \cos^{-1}\left(\frac{-10}{\sqrt{500}}\right)$$

$$\angle ABC = 116 \cdot 6°$$

(iii) Triangular concurrency facts can be established by vector methods.
This has never come up in an exam but given that it is mentioned in the SQA conditions and arrangements, it could.

Example

Consider the triangle ABC with median AM.

The point G cuts AM in the ratio 2:1. Express *g*, the position vector of G in terms of *a*, *b* and *c* the position vectors of A, B and C respectively.

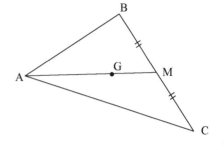

Let *m* be the position vector of M, the midpoint of BC thus $m = {}^{(b + c)}\!/_2$

$$\frac{\overrightarrow{AG}}{\overrightarrow{GM}} = \frac{2}{1}$$

$\Rightarrow g - a = 2(m - g)$

$\Rightarrow g = 2m - 2g + a$

$\Rightarrow 3g = b + c + a$

$\Rightarrow g = (a + b + c)/3$

Similarly if we try to find G′, the point that divides the median BN in the ratio 2:1, where N is the midpoint of AC we get …

Let *n* be the position vector of M, the midpoint of BC, thus $n = {}^{(a + c)}\!/_2$

$$\frac{\overrightarrow{BG'}}{\overrightarrow{G'N}} = \frac{2}{1}$$

$\Rightarrow g' - b = 2(n - g')$

$\Rightarrow g' = 2n - 2g' + b$

$\Rightarrow 3g' = a + c + b$

$\Rightarrow g' = (a + b + c)/3$

In other words G and G′ are the same point.

The same result would ensue if we tried to find the point on the median CQ which divided it in the ratio 2:1.

Thus the three medians are concurrent (all run through the same point) and the point of concurrency is also the point of trisection of each median.

This point G is called the centroid of the triangle.

Objective questions

❶   A is the point (1, 5, 3) and AB = $\begin{pmatrix} 1 \\ 3 \\ -2 \end{pmatrix}$.

What are the coordinates of the point B?

A (2, 8, 1)   B (0, −2, −5)   C (0, 2, 5)   D (2, 8, 5)

Rough working

$\vec{AB} = \mathbf{b} - \mathbf{a}$

$$\mathbf{b} = \vec{AB} + \mathbf{a} = \begin{pmatrix} 1 \\ 3 \\ -2 \end{pmatrix} + \begin{pmatrix} 1 \\ 5 \\ 3 \end{pmatrix} = \begin{pmatrix} 2 \\ 8 \\ 1 \end{pmatrix}$$

Choose option A

❷ What is the size of the angle between the vectors $\begin{pmatrix} 1 \\ 1 \\ \sqrt{2} \end{pmatrix}$ and $\begin{pmatrix} 1 \\ 0 \\ 0 \end{pmatrix}$?

A π   B π/2   C π/3   D π/4

Rough working

$$\begin{pmatrix} 1 \\ 1 \\ \sqrt{2} \end{pmatrix} \begin{pmatrix} 1 \\ 0 \\ 0 \end{pmatrix} = 1$$

$$\left\| \begin{pmatrix} 1 \\ 1 \\ \sqrt{2} \end{pmatrix} \right\| = \sqrt{1^2 + 1^2 + (\sqrt{2})^2} = 2$$

$$\left\| \begin{pmatrix} 1 \\ 0 \\ 0 \end{pmatrix} \right\| = 1$$

$\cos \theta = \frac{1}{2}$

$\Rightarrow \theta = \frac{\pi}{3}$

Choose option C

❸ The vectors $3\mathbf{i} + 2\mathbf{j} - 2\mathbf{k}$ and $2\mathbf{i} + x\mathbf{j} + 2x\mathbf{k}$ are mutually perpendicular. What is the value of x?

A 1   B 2   C 3   D 4

Rough working

Scalar product = 0, so $6 + 2x - 4x = 0$.
So $x = 3$.

Choose option C

4 The edges of a cuboid represent vectors **a**, **b** and **c** as shown.

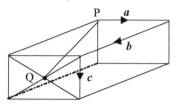

P is a vertex of the cuboid and Q is the centre of one of the faces as shown.

Express the directed line segment $\overrightarrow{PQ}$ in terms of **a**, **b** and **c**.

A $a + b + ½$
B $½ + b + ½$
C $½ + b + c$
D $(a + b + c)/2$

Rough working

To get from P to Q we can 'walk' half way along **a**, along **b**, and halfway down **c**

$½ + b + ½$.

Choose option B

5 The points A(1, 2, 3), B(x, 5, z) and C(13, 11, −12) are collinear.
What are the values of x and z?

A $x = 5; z = 2$
B $x = 5; z = -2$
C $x = -5; z = 2$
D $x = -5; z = -2$

Rough working

If the points are collinear then $\overrightarrow{AB} = k\overrightarrow{AC}$ for some $k \in R$

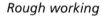

$$\overrightarrow{AB} = b - a = \begin{pmatrix} x - 1 \\ 5 - 2 \\ z - 3 \end{pmatrix} = \begin{pmatrix} x - 1 \\ 3 \\ z - 3 \end{pmatrix}$$

$$k\overrightarrow{AC} = k(c - a) = k\begin{pmatrix} 13 - 1 \\ 11 - 2 \\ -12 - 3 \end{pmatrix} = \begin{pmatrix} 12k \\ 9k \\ -15k \end{pmatrix}$$

Comparing components we see that
(a) $9k = 3 \Rightarrow k = \frac{1}{3}$
(b) $12k = x - 1 \Rightarrow 4 = x - 1 \Rightarrow x = 5$
(c) $-15k = z - 3 \Rightarrow -5 = z - 3 \Rightarrow z = -2$

Choose option B

In this example we could just have substituted the options until we found a pair of values that gave us $\overrightarrow{AB} = k\overrightarrow{AC}$.

An extended response question that doesn't need a calculator

6 The diagram shows a crystal of the mineral beryl.
It is a hexagonal prism: *u*, *v* and *w* are unit vectors with $\overrightarrow{AB}$ representing *u*, $\overrightarrow{BC}$ representing *v* and $\overrightarrow{AG}$ representing *w*.

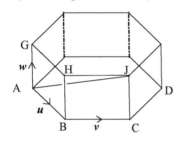

(a) Given that the hexagon is regular, express
    (i) $\overrightarrow{AC}$
    (ii) $\overrightarrow{AD}$ in terms of *u* and *v*.   *2 marks*
(b) Express $\overrightarrow{AJ}$ in terms of *u*, *v* and *w*.   *1 mark*
(c) Find the scalar product of $\overrightarrow{AD}$ with $\overrightarrow{AJ}$.   *3 marks*

(**Response**)────────────────────────────

(a) (i) $\overrightarrow{AC} = u + v$ (ii) $\overrightarrow{AD} = 2v$

Note: we get the second result by drawing the regular hexagon and its diagonals.
(b) $\overrightarrow{AJ} = u + v + w$
(c) $\overrightarrow{AD}.\overrightarrow{AJ} = 2v.(u + v + w)$
        $= 2v.u + 2v.v + 2v.w$
        $= 2|v||u| \cos 120° + 2|v||v| \cos 0°$
        $+ 2|v||w| \cos 90°$
        $= 2.1.1.-\frac{1}{2} + 2.1.1.1 + 2.1.1.0$
        $= -1 + 2 + 0$
        $= 1$

> ⚠ Remember, cos 120° = cos(180 − 60)° = −cos 60°.

Marking scheme

- •¹ interpret 2D representation of a 3D situation   *ic*
  [(a) (i)]
- •² take an economical approach, use symmetry   *ss*
  [(a) (ii)]
- •³ interpret 2D representation of a 3D situation   *ic*
  [(b)]
- •⁴ identify crucial aspect   *ss*
  [to see scalar product is commutative, i.e. we
  can get rid of brackets in this fashion]
- •⁵ identify crucial aspect   *ss*
  [to identify correct angles]
- •⁶ evaluate a scalar product   *pd*

Note that the C-grade skills lie in parts (a) and (b).
The level of complexity would push part (c) into A/B grade territory.

An extended response question that needs a calculator

❼ In an experiment, a beam of light propagated from a point A strikes the surface of a beaker of water at B, bends as it enters the water and strikes the bottom of the beaker at C. Relative to a suitable set of axes the points A, B and C are measured as A(3, 5, 12), B(1, 2, 8) and C(−2, 1, 1).

Through what angle does the light bend?   *6 marks*

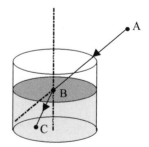

The context here might be familiar to students of physics.

To use the scalar product we want two vectors that are tail-to-tail.
The ray of light should have gone straight on along the dotted line but it bent towards the vertical axis. The angle bent through is the angle between the dotted extention to AB and BC. So we want to concentrate on the scalar product $\overrightarrow{AB}.\overrightarrow{BC}$ and not $\overrightarrow{BA}.\overrightarrow{BC}$ (e.g. using the distance formula and trigonometry) we could use the scalar product.

$$\overrightarrow{AB} = \mathbf{b} - \mathbf{a} = \begin{pmatrix} 1-3 \\ 2-5 \\ 8-12 \end{pmatrix} = \begin{pmatrix} -2 \\ -3 \\ -4 \end{pmatrix}$$

$$\overrightarrow{BC} = \mathbf{c} - \mathbf{b} = \begin{pmatrix} -2-1 \\ 1-2 \\ 1-8 \end{pmatrix} = \begin{pmatrix} -3 \\ -1 \\ -7 \end{pmatrix}$$

$$\overrightarrow{AB}.\overrightarrow{BC} = \begin{pmatrix} -2 \\ -3 \\ -4 \end{pmatrix}\begin{pmatrix} -3 \\ -1 \\ -7 \end{pmatrix} = 6 + 3 + 28 = 37$$

$$|\overrightarrow{AB}| = \sqrt{(-2)^2 + (-3)^2 + (-4)^2} = \sqrt{29}$$

$$|\overrightarrow{BC}| = \sqrt{(-3)^2 + (-1)^2 + (-7)^2} = \sqrt{59}$$

$$\cos x = \frac{37}{\sqrt{1711}}$$

Required angle = 26·6° (to 3 s.f.)

Each step of the strategy is exposed mainly to improve the chances of capturing marks. It also improves your chances of gaining follow-through marks if you make an error. Follow-through marks are given when, even if you've made an error in an earlier part of the question which means that your final answer is wrong, your working is consistent.

Marking scheme

- ●1 interpret 2D representation of a 3D situation   *ic*
  [identify which angle is required]
- ●2 strategy to use, $\cos\theta = \dfrac{\mathbf{a}.\mathbf{b}}{|\mathbf{a}||\mathbf{b}|}$ *ss*
- ●3 interpret scalar product $a_1 b_1 + a_2 b_2 + a_3 b_3$ *ic*
  [evaluated for the mark]
- ●4 get the magnitude of AB   *pd*
- ●5 get the magnitude of BC   *pd*
- ●6 evaluate angle   *pd*

In this marking scheme, if you fail to identify the correct angle but go on to calculate the size of ∠ABC say, you can still achieve five out of six marks... but only if it's totally clear that you know the steps. That's why your working should be explicit and transparent.

An extended response question that doesn't need a calculator

A balancing toy pivots at A(3, 2, 5). It has a counterweight at B(7, 4, 9).
At point C there is a little hook.
C divides AB internally in the ratio 3:1
(a) What are the coordinates of C?   *3 marks*
(b) A toy model plane is positioned on the balancing beam at a
    point P which divides AB *externally* in the ratio 3:5. What are
    the coordinates of P?   *3 marks*

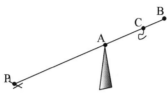

(**Response**)

Note that part (a) uses C-grade skills and part (b) uses A/B grade skills.

(a) $\dfrac{\overrightarrow{AC}}{\overrightarrow{CB}} = \dfrac{3}{1}$

$\Rightarrow c - a = 3b - 3c$
$\Rightarrow 4c = a + 3b$
$\Rightarrow 4c = 3i + 2j + 5k + 3(7i + 4j + 9k)$
$\Rightarrow 4c = 24i + 14j + 32k$
$\Rightarrow c = 6i + 3·5j + 8k$

C is the point (6, 3·5, 8) [This could have been written out using
components instead].

(b) $\dfrac{\overrightarrow{AP}}{\overrightarrow{PB}} = \dfrac{-3}{5}$

$\Rightarrow 5p - 5a = -3b + 3p$
$\Rightarrow 2p = 5a - 3b$
$\Rightarrow 2p = 15i + 10j + 25k - 21i - 12j - 27k)$
$\Rightarrow 2p = -6i - 2j - 2k$
$\Rightarrow p = -3i - 1j - 1k$

P is the point (−3, −1, −1).

Marking scheme

●^{1} interpret the expression 'internally in ratio'　　　*ic*

[giving $\dfrac{\overrightarrow{AC}}{\overrightarrow{CB}} = \dfrac{3}{1}$]

●^{2} determine the coordinates of point which
divide internally in a given ratio　　　*ss*

●^{3} manipulate vectors to solution　　　*pd*
[evaluated for the mark]

●^{4} interpret the expression 'externally in ratio'　　　*ic*

[$\dfrac{\overrightarrow{AP}}{\overrightarrow{PB}} = \dfrac{-3}{5}$]

●^{5} determine the coordinates of point which
divide externally in a given ratio　　　*ss*

●^{6} manipulate vectors to solution　　　*pd*

Note that marks 4, 5 and 6 are deemed grade A/B.

An A/B short response question

Find the unit vector, u, which is parallel to the vector $v = 2i + 3j + 6k$.

Response

u is parallel to $v \Rightarrow ku = v$ and $k|u| = |v|$
u is a unit vector $\Rightarrow |u| = 1$
$\Rightarrow k = |v| = \sqrt{(2^2 + 3^2 + 6^2)} = 7$
$\Rightarrow 7u = v = 2i + 3j + 6k$
$\Rightarrow u = \frac{1}{7}v = \frac{2}{7}i + \frac{3}{7}j + \frac{6}{7}k$

Marking scheme

●^{1} interpret a collection of facts before proceeding with task　　　*ic*
(unit vector, parallel ... $|u| = 1$; $ku = v$)

●^{2} marshal facts and make deductions　　　*ss*
$k|u| = |v|$; $k = |v|$

●^{3} evaluate formula (find magnitude of v)　　　*pd*

●^{4} establish expression for u　　　*pd*

It is mark 2 which pulls this question into the A/B category. For level C you would be expected marshal facts and make deductions but only up to about three steps.

Topic Tips

The most commonly asked question in this topic involves finding an angle in space, given the coordinates of the vertex and a point on each arm. It tests lots of skills. Make sure you are familiar with this type of question. Vector walks are commonly overlooked by students as being too easy and so, paradoxically, are badly done when they come up. The main A/B differentiator in this topic is the strategy for determining the position vector of a point that divides a line **externally** in a given ratio. It would be a pity to miss out on these marks.

13 | Further Differentiation and Integration

What you should know

Related problems

Objective questions

Extended response questions that need a calculator

A short response question that needs a calculator

An extended response question that doesn't need a calculator

What you should know

You are expected to know the following facts.

- If $f(x) = \sin x$, then $f'(x) = \cos x$ and $\int \cos x \ dx = \sin x + c$.
- If $f(x) = \cos x$, then $f'(x) = -\sin x$ and $\int \sin x \ dx = -\cos x + c$.
- If $f(x) = g(h(x))$, then $f'(x) = g'(h(x)).h'(x)$.
  This is referred to as the chain rule.
- Integrate functions defined by $f(x) = (px + q)^n$ for all rational n, except $n = -1$.
- Integrate functions defined by $f(x) = p \cos (qx + r)$ and $f(x) = p \sin (qx + r)$.

Related problems

You should be able to use this knowledge to solve the following kinds of problems.

- Differentiating composite functions.
  A verbal mnemonic is often used when applying the chain rule.
  One has to imagine an 'outside' function and an 'inside' function.
  'Differentiate *outside* and multiply by the derivative of the *inside*.'

Example 1
Differentiate $y = (3x^2 + 4x + 1)^3$

(Response) ───

Here the outside function is 'something' cubed: the inside function is $3x^2 + 4x + 1$

$\dfrac{dy}{dx} = 3(3x^2 + 4x + 1)^2(6x + 4)$ ─────────────────── ⊂⊃

Example 2
Differentiate $y = \sin^3 x$ ... [level A/B]

(Response) ───

> **! .** Remember that $\sin^3 x$ is an alternative way of writing $(\sin x)^3$.

... the outside function is 'something' cubed: the inside function is $\sin x$

$\dfrac{dy}{dx} = 3\sin^2 x.\cos x$ ───────────────── ⊂⊃

- Integrating composite functions when the 'inside' function is linear.

> **! .** If $\int f(x)\ dx = F(x) + c$ then $\int f(ax + b)\ dx = \frac{1}{a} F(ax + b) + c$.

Example 1
We know $\int \cos x\ dx = \sin x + c$ so we can deduce that

$\int \cos(3x + 5)\ dx = \frac{1}{3}\sin(3x + 5) + c$

Example 2
$\int \dfrac{1}{(3x + 4)^5} dx$ ... [level A/B]

This can be rewritten as $\int (3x + 4)^{-5}\ dx = \frac{1}{3}\left(\dfrac{(3x + 4)^{-4}}{-4}\right) + c$

Do not try to simplify the expression. As it stands, you can see all the parts of the strategy.

- Integrate the sum or difference of such functions, where p, q and r are constants.
  Here you will be examined on the knowledge that the integral of a sum is the sum of the integrals of the individual terms.
  i.e. $\int f(x) + g(x)\ dx = \int f(x)\ dx + \int g(x)\ dx$

A table of the standard trigonometric derivatives and integrals will be given in the Formulae List in the exam.

$f(x)$	$f'(x)$
$\sin ax$	$a\cos ax$
$\cos ax$	$-a\sin ax$

and

$f(x)$	$\int f(x)\, dx$
$\sin ax$	$-\dfrac{1}{a}\cos ax + c$
$\cos ax$	$\dfrac{1}{a}\sin ax + c$

[See appendix 4 for the complete list.]

- Candidates should be familiar with other forms of the chain rule such as
$$\frac{dy}{dx} = \frac{dy}{du} \times \frac{du}{dx}$$

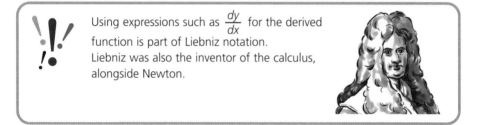

Using expressions such as $\dfrac{dy}{dx}$ for the derived function is part of Liebniz notation.
Liebniz was also the inventor of the calculus, alongside Newton.

There are a few places where this might crop up at Higher level. The examiners may specifically ask for its use.

Example

$$y = \frac{3}{(x^2 + 3x + 1)^5}$$

(a) If $u = x^2 + 3x + 1$. Find an expression for $\dfrac{du}{dx}$.

(b) Find an expression for $\dfrac{dy}{du}$.

(c) Hence find an expression for $\dfrac{dy}{dx}$.

(a) $u = x^2 + 3x + 1 \Rightarrow \dfrac{du}{dx} = 2x + 3$

(b) $y = \dfrac{3}{(x^2 + 3x + 1)^5} \Rightarrow y = \dfrac{3}{u^5} = 3u^{-5} \Rightarrow \dfrac{dy}{du} = -15u^{-6}$

(c) $\dfrac{dy}{dx} = \dfrac{dy}{du} \times \dfrac{du}{dx} = (-15u^{-6})(2x + 3) = (-15(x^2 + 3x + 1)^{-6})(2x + 3)$

$\quad = \dfrac{-15(2x + 3)}{(x^2 + 3x + 1)^6}$

It is interesting to note that differentiation with respect to x is considered grade C but differentiation with respect to any other variable would be considered grade A/B. Thus the above example would be classified as grade A/B since we have to differentiate with respect to u.

In a similar vein when we use differentiation in the context of rates of change, especially speed and displacement, we might enter into this category of problem.

Example
A rocket sledge moves away from its point of origin such that $s = 10t^2 + 5t$ while the fuel is burning where s is the distance from the start in metres and t is the time in seconds since the burn was started.

(a) How far has the sledge travelled after three seconds?
(b) What is its speed after four seconds?
(c) When the fuel is spent the sledge has attained a speed of 125 m/s.
   For how long is the fuel burning?

Response

(a) $t = 3 \Rightarrow s = 10 \times 3^2 + 5 \times 3 = 105$ metres
   This is worth one mark – the question establishes that the candidate has interpreted the question's context. ... a grade C mark.

(b) $v = \dfrac{ds}{dt} = 20t + 5$ so at $t = 4$, $v = 85$ m/s
   This is a grade A/B question because the candidate differentiates with respect to t.

(c) $20t + 5 = 125 \Rightarrow t = 6$ seconds
   This is a fairly easy interpretation mark... but it is unavailable to C-grade candidates because they will be unable to bridge the gap between part (a) and part (c).

The Leibniz form of the chain rule may also crop up when dealing with related rates of change.

'A particle moves along the x-axis at 3 m/s in the positive direction.' This can be written as '$\dfrac{dx}{dt} = 3$ m/s' where t is the time in seconds.

Example

A spherical bubble grows so that its radius, x mm, is increasing at 2 mm/s.

How fast is its volume increasing in mm³/s when the radius is 20 mm?

(*Response*)

The first statement tells us that $\frac{dx}{dt} = 2$ mm/s.

We know a formula for the volume of a sphere, $V = \frac{4}{3}\pi x^3$.

So we can work out that $\frac{dV}{dx} = 4\pi x^2$.

Now using the chain rule: $\frac{dV}{dt} = \frac{dV}{dx} \times \frac{dx}{dt}$

$\Rightarrow \frac{dV}{dt} = 4\pi x^2 \times 2 = 8\pi x^2$.

When $x = 20$ we can calculate the volume is increasing at $8 \times \pi \times 20^2 = 10\ 053$ mm³/s to the nearest whole number.

- You should be able to differentiate and integrate a variety of composite functions.

  The SQA specifies types of functions that they would consider grade C and those they would consider grade A/B.

  Grade C: A simple power of a linear function where the coefficient of x is 1.

  Examples
  (a) Differentiate $(x + 4)^{\frac{1}{2}}$ *Response* $\frac{1}{2}(x + 4)^{-\frac{1}{2}}$
  (b) Differentiate $(x - 2)^{-3}$ *Response* $-3(x - 2)^{-4}$

  Grade A/B: A simple power or trig function of a linear function where the coefficient of x is not 1.

  Examples
  (a) Differentiate $(2x - 5)^3$ *Response* $3(2x - 5)^2 \times 2 = 6(2x - 5)^2$
  (b) Differentiate $\sin 3x$ *Response* $\cos 3x \times 3 = 3\cos 3x$
  (c) Differentiate $\cos 3x$ *Response* $-\sin 3x \times 3 = -3\sin 3x$

  Integration also follows similar grading.

  Example
  Grade C
  (a) Integrate $(x + 4)^{\frac{1}{2}}$ *Response* $\frac{2}{3}(x + 4)^{\frac{3}{2}} + c$
  (b) Integrate $3\cos x$.   *Response* $3\sin x + c$

  Grade A/B
  (a) Integrate $(3x + 1)^3$ *Response* $\frac{1}{4}(3x + 1)^4 \times \frac{1}{3} + c$
                                     $= \frac{1}{12}(3x + 1)^4 + c$
  (b) Integrate $\sin 2x$ *Response* $-\frac{1}{2}\cos 2x + c$

- Apart from the new range of functions which the candidate can now tackle, he/she will be expected to solve the same sort of problems learned in calculus in unit 1 and unit 2 namely.

  In differential calculus:
  (i) finding the equation of a tangent to a curve;
  (ii) finding if/when a function is increasing/decreasing/stationary;
  (iii) finding rates of change;
  (iv) finding maxima and minima.

  In integral calculus:
  (i) finding the anti-derivative;
  (ii) solving simple differential equations;
  (iii) finding the area under a curve.

Objective questions

 1 Differentiate $\sin(3x + 2)$.

    A $3\cos(3x + 2)$
    B $-3\cos(3x + 2)$
    C $\frac{1}{3}\cos(3x + 2)$
    D $-\frac{1}{3}\cos(3x + 2)$

Rough working

The derivative of sin is cos ... so ignore the negatives.
The derivative of the inside function is 3 ... ignore the $\frac{1}{3}$.

Choose option A

2 Differentiate $\sin(2 - 3x)$.

    A $3\cos(2 - 3x)$
    B $-3\cos(2 - 3x)$
    C $\frac{1}{3}\cos(2 - 3x)$
    D $-\frac{1}{3}\cos(2 - 3x)$

Rough working

This is an A/B grade question.
The derivative of sin is cos ... $\cos(2 - 3x)$.
The derivative of the inside function is -3 ... careful here!
The answer is $-3\cos(2 - 3x)$.

Choose option B

❸ What is the gradient of the tangent to the curve $y = 2\sin x$ at the point where $x = \pi/3$?

A $\sqrt{3}/2$

B ½

C 1

D $\sqrt{3}$

Rough working

$\dfrac{dy}{dx} = 2\cos x$

$x = \frac{\pi}{3} \Rightarrow$ gradient is $2\cos \frac{\pi}{3} = 2 \times \frac{1}{2} = 1$

Choose option C

❹ What is the equation of the tangent to the curve $y = (x - 3)^7$ at the point where $x = 2$?

A $y = 7x - 15$

B $y = 7x - 1$

C $y + 7x = 15$

D $y + 7x = -15$

Rough working

$\dfrac{dy}{dx} = 7(x - 3)^6$

$x = 2 \Rightarrow y = -1$ and $\dfrac{dy}{dx} = 7$

Equation is thus: $y + 1 = 7(x - 2) \Rightarrow y = 7x - 15$.

Choose option A

❺ A particle moves so that its distance from the origin, y cm can be calculated from the formula $y = \sin 2x$ where x is the time in seconds since the start of observations.

What would be its *velocity* in cm/sec after $\pi/6$ seconds?

A $\sqrt{3}/2$ cm/s

B ½ cm/s

C $\sqrt{3}$ cm/s

D 1 cm/s

Rough working

Because of the presence of the double angle in differentiation the question is automatically grade A/B.

Velocity, v, is the rate of change of displacement with time.

$\dfrac{dy}{dx} = 2\cos 2x$

when $x = \frac{\pi}{6}$, $v = 2 \cos \left(\frac{\pi}{3}\right) = 2 \times \frac{1}{2} = 1$

Choose option D

6 The function $f(x) = \dfrac{1}{(x + 1)^5}$, $x \geq 0$ fits which of the following descriptions?

A   It's always increasing.
B   It's always decreasing.
C   It's either increasing or stationary.
D   It's either decreasing or stationary.

Rough working

$\dfrac{dy}{dx} = \dfrac{-5}{(x + 1)^6}$ ... any non-zero quantity to the power 6 is always positive.

So the derivative is always negative.
So the function is always decreasing.
Note the derivative can't be zero.

Choose option B

7 The function $f(x) = x + \sin x$ fits which of the following descriptions?

A   It's always increasing.
B   It's always decreasing.
C   It's either increasing or stationary.
D   It's either decreasing or stationary.

Rough working

$\dfrac{dy}{dx} = 1 + \cos x$
$-1 \leq \cos x \leq 1$
$\Rightarrow 0 \leq 1 + \cos x \leq 2$
So derivative is either zero (stationary point) or positive (increasing function).

Choose option C

8 What is the area trapped between $y = \sin x$ and $y = \cos x$ in the interval $0 \leq x \leq \pi/2$?

A $2\sqrt{2} - 2$
B $2\sqrt{2} + 2$
C $2 - \sqrt{2}$
D 1

Rough working

area $= 2\displaystyle\int_0^{\pi/4} \cos x - \sin x \; dx = 2[\sin x + \cos x]_0^{\pi/4}$

$2[(1/\sqrt{2} + 1/\sqrt{2}) - (0 + 1)] = 2[\sqrt{2} - 1] = 2\sqrt{2} - 2$

Choose option A

⑨ Integrate $(3 - 2x)^5$

   A $5(3 - 2x)^4 + c$
   B $-10(3 - 2x)^4 + c$
   C $\frac{1}{6}(3 - 2x)^6 + c$
   D $-\frac{1}{12}(3 - 2x)^6 + c$

Rough working

Note that the coefficient of x in the 'inside' function $\neq 1$ – so we have a grade A/B question.

Add one to power and divide by new power ... $\frac{1}{6}(3 - 2x)^6$

... **then** divide by coefficient of x ... $-\frac{1}{12}(3 - 2x)^6$.

Choose option D

⑩ $\displaystyle\int_{\pi/6}^{\pi/3} \sin 3x \; dx$ is equal to:

   A   1
   B -1
   C $\frac{1}{3}$
   D $^-\frac{1}{3}$

Rough working

Another level A/B question.

$$\left[-\frac{1}{3}\cos 3x\right]_{\pi/6}^{\pi/3} = -\tfrac{1}{3}\cos \pi + \tfrac{1}{3}\cos \tfrac{\pi}{2} = -\tfrac{1}{3}.-1 + \tfrac{1}{3}.0 = \frac{1}{3}$$

Choose option C

⑪ The derivative of a function is $f(x) = \cos(x - 1)$ and $f(1) = 1$.
Find a formula for $f(x)$.

   A $f(x) = \sin(x - 1)$
   B $f(x) = \sin(x - 1) - 1$
   C $f(x) = 1 - \sin(x - 1)$
   D $f(x) = \sin(x - 1) + 1$

Rough working

Integrate to find $f(x)$ ... $\sin(x - 1) + c$.
When $x = 1$, $y = 1$ ... $1 = \sin 0 + c \Rightarrow c = 1$
$f(x) = \sin(x - 1) + 1$.

Choose option D

⑫ What is the derivative of sin $x°$?

A $\cos x°$

B $\dfrac{\pi}{180}\cos x°$

C $\dfrac{180}{\pi}\cos x°$

D there is no derivative when working in degrees.

Rough working

This question, by its working, will turn out to be grade A/B.

 $1° = \frac{\pi}{180}$ radians

Turn the angle into radians by multiplying by $\frac{\pi}{180}$
$y = \sin(\frac{\pi}{180}x)$
Using chain rule, the derivative $= \frac{\pi}{180}\cos(\frac{\pi}{180}x)$ ... this is the line which makes it A/B.
Turn the angle into degrees by multiplying by $\frac{180}{\pi}$
$= \frac{\pi}{180}\cos(x°)$

Choose option B

Note: this is the sort of trivial pursuit you should learn anyway. Being conscious of it keeps you straight ... it's a reminder that when working in calculus you should work in radians.

⑬ A particle moves along the x-axis so that its velocity v m/s is given by
$v = 3 + \sin x$
where x is the time in seconds since the start of observations.
Find a formula for the distance from the origin, s metres assuming the particle was 1 metre away when the observations began.

A $s = 3x - \cos x$

B $s = -\cos x - 1$

C $s = 3x - \cos x + 2$

D $s = \cos x + 1$

Rough working

We differentiate displacement to get velocity ... so integrate velocity to get displacement ... but don't forget c, the constant of integration.

$s = \int 3 + \sin x \ dx = 3x - \cos x + c$

When $x = 0$ displacement $= 1$

$1 = 0 - 1 + c \Rightarrow c = 2$

$s = 3x - \cos x + 2$

Choose option C

Extended response questions that need a calculator

⑭ A man has a rowing boat. He can travel on the river at 6 km/h. He can walk along the river bank at 8 km/h. The river is a kilometre wide. He wishes to cross the river and make his way down to a house 4 km away on the other bank (see sketch).

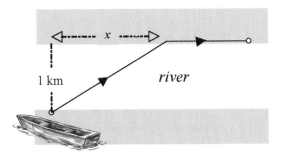

(a) Show that the time, T hours, it would take him is given by the formula

$T = \dfrac{\sqrt{1 + x^2}}{6} + \dfrac{4 - x}{8}$ where x km is the distance along the

bank where he lands the boat. *3 marks*

(b) Calculate the value of x which would minimise this time. *7 marks*

(Response) ───

This format is a fairly common occurrence. You are being asked to establish a relationship and then to use it to find some critical condition. It is also fairly common that a candidate cannot find how to produce the formula.

This should not stop you from using the formula to find the required condition.

(a) By Pythagoras' Theorem, we see the distance travelled by boat is $\sqrt{(1 + x^2)}$.

Using $T = D \div S$ this leg of the journey takes $\sqrt{(1 + x^2)}/6$.

The distance travelled on land $= 4 - x$ (simple subtraction).

Using $T = D \div S$ this leg of the journey takes $(4 - x)/8$.

Adding the two times gives the required result.

(b) $T = \dfrac{1}{6}(1 + x^2)^{1/2} + \dfrac{4}{8} - \dfrac{x}{8}$

$\dfrac{dT}{dx} = \dfrac{1}{12}(1 + x^2)^{-1/2}.2x - \dfrac{1}{8}$

$= 0$ at stationary points.

$\Rightarrow \dfrac{x}{6(1 + x^2)^{1/2}} = \dfrac{1}{8}$

$\Rightarrow \dfrac{x^2}{1 + x^2} = \dfrac{36}{64}$

$\Rightarrow 64x^2 = 36 + 36x^2$

$\Rightarrow 28x^2 = 36$

$\Rightarrow x = \sqrt{(36/28)} = \sqrt[3]{7}$

$\Rightarrow x = 1\cdot134$ (to 3 d.p.)

The man should aim for a point 1134 metres downstream to minimise his journey.

A table should accompany this answer showing that you have explored the nature of the stationary point that you found.
You have to explore the sign of the derivative around the stationary point. The following would be considered the minimum response in order to gain the mark.

x	$\rightarrow$	$\sqrt[3]{7}$	$\rightarrow$
$\dfrac{dy}{dx}$	–	0	+
behaviour	╲	—	╱

The function has a minimum turning point at $x = \sqrt[3]{7}$.
NOW check you have answered the question.
Has the minimum time been requested? In this case, no.

Marking scheme

•^{1}	strategy for distances	ss
•^{2}	strategy for times	ss
•^{3}	processing the algebra	pd
•^{4}	know to differentiate	ss
•^{5}	use chain rule	pd
•^{6}	differentiate	pd
•^{7}	set derivative to zero and start solving	ss
•^{8}	solve for x	pd
•^{9}	explore nature of T.P.	ic
•^{10}	express solution in language appropriate to the situation	ic

Sometimes solving problems of this type throws up more than one answer. There will be a communication mark for saying why any value found is discarded.

Answers to questions like this can often be found by exploring with a calculator ... but you won't get any marks.

Plotting the function into a graphic calculator will also help you find minima and maxima.

However, if you find the answer this way it will be assumed that it's been obtained from a scale drawing and once again will attract no marks. It could be a good way to check if your answer obtained from an analytical approach is correct.

The SQA do not ask trick questions. So if they ask you to find a minimum value there will be one. This does not mean that the value you found *is* the minimum value. You still have to check the nature by means of a table of signs.

You only need a calculator to gain mark 10.

⑮ The depth of water, *h* metres, in a small harbour can be modeled by the function

$h(x) = \cos(\tfrac{\pi}{6}x + \tfrac{2}{3}) + 3$ where *x* is the time in hours since midnight.

(a) How deep is the water at 4pm?

(b) What is the maximum depth attained in the harbour between 12 noon and 7pm?

(Response)

(a) At 4pm $x = 16$

$h(16) = \cos(\tfrac{16\pi}{6} + \tfrac{2}{3}) + 3 = 2$

Harbour is 2 metres deep at 4pm.

(b) In a closed interval the minimum will occur at either an end-point or a turning point. So we must examine **all** these points.

$h'(x) = -\tfrac{\pi}{6}\sin(\tfrac{\pi x}{6} + \tfrac{2}{3})$

$\qquad = 0$ at stationary points

$\Rightarrow \tfrac{\pi x}{6} + \tfrac{2}{3} = ..., 0, \pi, 2\pi, ...$

$\Rightarrow \tfrac{\pi x}{6} = -\tfrac{2}{3}, \tfrac{2\pi}{3}, \tfrac{5\pi}{3}, \tfrac{8\pi}{3}, \tfrac{11\pi}{3}, ...$

$\Rightarrow x = -2, 4, 10, 16, 22, ...$

We are only interested in the interval $12 \le x \le 19$.

The only turning point in this interval is at $x = 16$

$h(16) = \cos(\tfrac{16\pi}{6} + \tfrac{2}{3}) + 3 = 2$.

Examine end-points too.

$h(12) = \cos(^{12}\!/_6 + {}^{\pi}\!/_3) + 3 = 3.5$

$h(19) = \cos(^{19}\!/_6 + {}^{\pi}\!/_3) + 3 = 3$

Since the maximum must occur at one of these three points then it must occur at 12 noon.

The maximum depth in this period is 3·5 metres which occurs at 12 noon.

Marking scheme

- •^{1} interpret equation — ic
- •^{2} strategy for finding turning points — ss
- •^{3} differentiate — pd
- •^{4} set derivative to zero and start solving — ss
- •^{5} solving trig equation — pd
- •^{6} identifying pertinent solution — ic
- •^{7} consider end-points of closed interval — pd
- •^{8} communicate minimum value in appropriate language — ic

A short response question that needs a calculator

16 Evaluate $\displaystyle\int_0^5 2x + 12\sin 3x \; dx$ *4 marks*

Response

$\displaystyle\int_0^5 2x + 12\sin 3x \; dx = \left[x^2 - {}^{12}\!/_3\cos 3x\right]_0^5$

$= (5^2 - 4\cos 15) - (0^2 - 4\cos 0)$

$= 28.039 + 4$

$= 32.0$ (to 3 s.f.)

Marking scheme

- •^{1} integral of $\sin x = -\cos x$ AND x^2 — pd
- •^{2} apply the chain rule — pd
  [evidence of 1/3]
- •^{3} handle limits — ss
  [substitution correct]
- •^{4} evaluate definite integral — pd

Note that a candidate who gets 25·1 from an otherwise correct response has been working in degrees and will lose a mark.

There's no need for the examiner to say that x is in radians. The context, calculus, is all the warning you need ... that and the absence of any degree sign.

Mark 2 is an A/B mark. The other marks are all C-grade marks.

An extended response question that doesn't need a calculator

Here's an example from the SQA 1998 Paper 1 that illustrates several points.

⑰ A sketch of part of the graph of $y = \sin 2x$ is shown in the diagram. The point P and Q have coordinates $(p, 0)$ and $(q, -1)$ respectively.

(a) Write down the values of p and q.

1 mark

(b) Find the area of the shaded region

4 marks

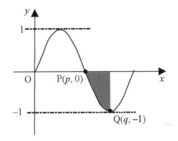

(Response) ───

(a) By inspection $p = \frac{\pi}{2}$ and $q = \frac{3\pi}{4}$.

You could have answered 90° and 135° and still have gained the mark. However, anticipating a strategy which involves calculus to find the area, it is far better to answer in radians.

(b) $\int_{\pi/2}^{3\pi/4} \sin 2x \ dx = \left[-\frac{1}{2}\cos 2x\right]_{\pi/2}^{3\pi/4}$

$= -\frac{1}{2}(\cos \frac{3\pi}{2} - \cos \pi) = -\frac{1}{2}(0 - (-1))$

$= -\frac{1}{2}$

The negative sign indicates that the area is below the x-axis. The shaded area is ½ square units. ────────────────( )

Marking scheme

●¹	interpret diagram and formula [to get value of both p and q]	*ic*
●²	know how to integrate	*ss*
●³	apply the chain rule	*pd*
●⁴	evaluate definite integral	*pd*
●⁵	interpret area below axis	*ic*

Although you would get the mark for saying $p = 90$, $q = 135$, you would lose a mark (mark 2) if you persisted in working in degrees.

The fifth mark is for interpreting the negative value for the definite integral and dealing with it. If you make a mistake (and lose a mark) and get a positive answer here, you cannot get the follow-through mark because you can't demonstrate your interpretation of a negative answer.

If you just drop the negative sign without making some remark then you will not be awarded mark 5.

For example

$= -\frac{1}{2}$

$= \frac{1}{2}$

is a fairly common response which does not earn mark 5.

Topic Tips

When revising this material make sure you pay particular attention to the chain rule, especially functions where the coefficient of x is not 1. These are sure to be A/B marks.

It is absolutely essential that you remember to work in radians when using calculus.

Remember, it's only in radians that the derivative of $\sin x$ is $\cos x$.

In degrees the derivative of $\sin x°$ is $\frac{\pi}{180} \cos x°$.

Marks get deducted when a student works in degrees when it's inappropriate.

14 Logarithmic and Exponential Functions

What you should know

Related problems

Objective questions

A short response question that doesn't need a calculator

A short response question that needs a calculator

An extended example that needs a calculator

What you should know

You are expected to know the following facts.

- $a^y = x \Leftrightarrow \log_a x = y$ $(a > 1, x > 0)$

  This is absolutely fundamental to your understanding of logs.
  It provides the definition of a log.
  It provides a way of changing the subject of a formula which involves logs.
  It provides the basic mechanism for solving related equations.
  It provides the tool for proving the laws of logs.
  Memorise it ... practise writing it on a regular basis and using different letters.
  $a^b = c \Leftrightarrow \log_a c = b$ ... note that in the switch from one form to the other, the base, a, remains the base.
  Before the advent of pocket calculators, people learned logs to make multiplication easier. Numbers were turned into logs (using a book of tables), then the logs were just added. Then this answer was turned back into a number (using the book of tables in reverse).
  We talked of logs and antilogs.
  $$y = \log_a x \Leftrightarrow x = \text{antilog}_a y$$
  We now say $y = \log_a x \Leftrightarrow x = a^y$
  ... maybe another way to remember it.

- The laws of logarithms:

$\log_a 1 = 0$ ... use the above and this becomes obvious ... $\log_a 1 = 0 \Leftrightarrow a^0 = 1$

$\log_a a = 1$ ... and the same here   ... $\log_a a = 1 \Leftrightarrow a^1 = a$

$\log_a(bc) = \log_a b + \log_a c$

Example $\log_{10}(2 \times 3) = \log_{10} 2 + \log_{10} 3$

It is often used the other way to simplify expressions.

Example $\log_{10} 2 + \log_{10} 5 = \log_{10}(2 \times 5) = \log_{10} 10 = 1$

$\log_a(b/c) = \log_a b - \log_a c$

Example $\log_{10}(\tfrac{2}{3}) = \log_{10} 2 - \log_{10} 3$

It also is often used the other way to simplify expressions.

Example $\log_{10} 8 - \log_{10} 4 = \log_{10}(8 \div 4) = \log_{10} 2$

$\log_a(b^n) = n\log_a b$

Example $\log_{10}(16) = \log_{10}(4^2) = 2\log_{10} 4$

This can be used to evaluate certain logs

Example $\log_5(125) = \log_5(5^3) = 3\log_5 5 = 3 \times 1 = 3$

And it's used the other way to simplify expressions.

Example $3\log_5 4 + 2\log_5 2 = \log_5 4^3 + \log_5 2^2 = \log_5 64 + \log_5 4 = \log_5 256$

$\log_a 1 = 0$
$\log_a a = 1$
$\log_a(bc) = \log_a b + \log_a c$
$\log_a(b/c) = \log_a b - \log_a c$
$\log_a(b^n) = n\log_a b$
$\log_a b = \log_a c \Leftrightarrow b = c$
$\log_a b > \log_a c \Leftrightarrow b > c$

It's handy to know that if $x > 0$ then $\log_a(\tfrac{1}{x}) = \log_a(x^{-1}) = -\log_a x$.
This has been the basis of more than one question in the past.

Example 1

Here is a 'proof' that $1 > 2$.
Follow the steps and say in which line the argument falls down.

$\dfrac{1}{2} > \dfrac{1}{4}$

$\Rightarrow \log_a(\tfrac{1}{2}) > \log_a(\tfrac{1}{4})$ ... taking the log of both sides

$\Rightarrow \log_a(\tfrac{1}{2}) > \log_a(\tfrac{1}{2})^2$ ... a half squared is a quarter

$\Rightarrow \log_a(\tfrac{1}{2}) > 2\log_a(\tfrac{1}{2})$ ... $\log a^b = b \log a$

$\Rightarrow 1 > 2$ ... dividing both sides by the same thing

Response

This argument brings out various things:
(i)   ½ *is* bigger than ¼
(ii) If $a > b$ then $\log a > \log b$
      ... this is true because the log
      function is an increasing function
      as the sketch shows
      ... so step 1 is legal.
(iii) $(½)^2 = ¼$
      ... so step 2 is legal.
(iv) $\log(½)^2 = 2 \log(½)$
      ... so step 3 is legal.

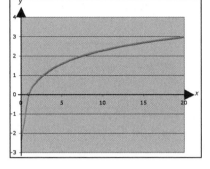

(v)   Step 4 is dividing throughout an inequation by $\log(½)$.
      But we know $\log(½) = - \log 2$ ... that is, it's negative (see sketch again).
      When we divide by a negative we must reverse the inequality.
(vi) So the last line should read $1 < 2$.

Example 2
The sketch shows the graph of the
function $y = \log_6 x$.
Make a sketch of the graph of
$y = \log_6 (¼)$.

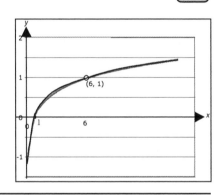

Response

$y = \log_6 (¼) = - \log_6 x$.

We need only reflect the graph
$y = \log_6 x$ in the x-axis to get what we
want.

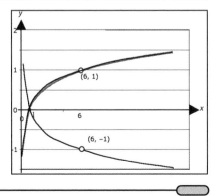

Related problems

You should be able to use this knowledge to solve the following kinds of problems.

- Simplify numerical expressions using the laws of logarithms.

Example 1
Simplify $\log_a 27 + \log_a 81$

Response

You should recognise powers of 3.
$\log_a 27 + \log_a 81 = \log_a 3^3 + \log_a 3^4 = 3\log_a 3 + 4\log_a 3 = 7\log_a 3$

Example 2
Simplify $\log_4 2 + \log_4(\tfrac{1}{4}) - \log_4 8$

Response 1

See them as powers of 4:
$\log_4 4^{\frac{1}{2}} + \log_4 4^{-1} - \log_4 4^{\frac{3}{2}}$
$= \tfrac{1}{2} \log_4 4 - \log_4 4 - \tfrac{3}{2} \log_4 4$
$= \tfrac{1}{2} - 1 - \tfrac{3}{2}$
$= -2$

Response 2

Use the laws of logs first:
$\log_4(2 \times \tfrac{1}{4} \div 8)$
$= \log_4(\tfrac{1}{16})$
$= -\log_4(16)$
$= -\log_4(4^2)$
$= -2 \log_4 4$
$= -2$

Example 3
$\log_e x + \log_e 3 = y$
Make x the subject of the formula.

Response

$\log_e x + \log_e 3 = y$
$\Rightarrow \log_e 3x = y$
$\Rightarrow 3x = e^y$
$\Rightarrow x = e^y/3$

The laws of logs have been used here to get the equation to the state $\log_a b = c$ ... which then allows us to switch to $b = a^c$.

> $\log_e x$ is an alternative notation for $\ln x$

> If you work with spreadsheets then EXP(x) is the same thing as e^x.
> = EXP(3) will return e^3 = 20 (to nearest whole number.)

- Solve simple logarithmic and exponential equations.
  At level C you should expect some guidance to get you 'into' the problem; otherwise these will be level A/B questions.

Example 1
Solve $\log_{10}x = 3.4$

(Response)

$\log_{10}x = 3.4 \Leftrightarrow x = 10^{3.4} \Leftrightarrow x = 2512$ to nearest whole number

Example 2
Solve $\log_{10}x + \log_{10}2x = 1$, $x > 0$

(Response)

$\log_{10}x + \log_{10}2x = 1$
$\Rightarrow \log_{10}2x^2 = 1$
$\Rightarrow 2x^2 = 10$
$\Rightarrow x = \sqrt{5}$

Example 3
Level A/B
Solve $\ln(2x + 1) = 1.4$

(Response)

$\ln(2x + 1) = 1.4$
$\Rightarrow 2x + 1 = e^{1.4}$
$\Rightarrow x = (e^{1.4} - 1)/2 = 1.53$ (3 s.f.)

Example 4, level A/B
£500 is placed in a bank account. At an interest rate of 3% per annum, the amount of money, £y in the bank account after x years is given by the formula $y = 500 \times 1.03^x$.
After how many complete years will there be more than £1000 in the account?

(Response)

We wish to solve the equation $1000 = 500 \times 1.03^x$.
$\Rightarrow 1.03^x = 1000 \div 500$
$\Rightarrow \ln(1.03^x) = \ln(2)$
$\Rightarrow x \ln(1.03) = \ln(2)$
$\Rightarrow x = \ln(2)/ \ln(1.03)$
$\Rightarrow x = 23.45$ (2 d.p)
$\Rightarrow$ after 24 years there will be more than £1000 in the bank.

Example 5
Simple examples of this type of equation can be level C.
Solve the equation $5 = 3^x$ correct to 3 significant figures.

(Response) ─────────────────────────────────

$5 = 3^x$
$\Rightarrow \ln(5) = \ln(3^x)$
$\Rightarrow x \ln(3) = \ln(5)$
$\Rightarrow x = \ln(5)/\ln(3) = 1{\cdot}46$

Example 6
Even the more complex version can be reduced to level C *where enough guidance is given in the question.*

By taking the logs of both sides, solve the equation $7.3^x = 2$.
[Note that the dot represents $\times$ and is not a decimal point.]

(Response) ─────────────────────────────────

$7.3^x = 2$
$\Rightarrow \ln(7.3^x) = \ln(2)$
$\Rightarrow \ln(3^x) + \ln(7) = \ln(2)$
$\Rightarrow x \ln(3) + \ln(7) = \ln(2)$
$\Rightarrow x = [\ln(2) - \ln(7)]/ \ln(3)$
$\Rightarrow x = -1{\cdot}14$

Here you have been given the strategy that makes
the question easier.

Example 7
Although not specifically mentioned in the conditions and arrangements, it is possible that the unknown is the base. This would make it an A/B question unless guidance is given.

Solve $3 = \log_x 5$

(Response) ─────────────────────────────────

$3 = \log_x 5$
$\Rightarrow x^3 = 5$
$\Rightarrow \ln(x^3) = \ln 5$
$\Rightarrow 3 \ln x = \ln 5$
$\Rightarrow \ln x = \ln 5 / 3 = 0{\cdot}5364793$
$\Rightarrow x = e^{0{\cdot}5364793} = 1{\cdot}71$ (to 3 s.f.)

- Solve for a and b given two pairs of corresponding values of x and y.

  There are three main types mentioned in the conditions and arrangements and each is A/B grade material.

  Type 1
  $\log y = a\log x + b$

  This will most commonly occur in the context of a graph.
  Having graphed $\log x$ against $\log y$ **a straight line** is obtained.
  The graph will thus have an equation of the form $\log y = m\log x + c$
  If we have two sets of data for x and y, then m and c can be found and its equation can be established.

  Example
  Two variables x and y are related by the equation $\ln y = m \ln x + c$ where m and c are constants. When $x = 2$ $y = 9$ and when $x = 5$ $y = 39$. Find m and c and write down the relationship between x and y.

  (Response)

  Using (2, 9): $\ln 9 = m \ln 2 + c$
  Using (5, 39): $\ln 39 = m \ln 5 + c$
  Subtracting: $\ln 39 - \ln 9 = m \ln 5 - m \ln 2 + c - c$
  $\Rightarrow \ln(39/9) = m \ln(5/2)$
  $\Rightarrow m = \ln(39/9) \div \ln(5/2) = 1\cdot60$ (3 s.f.)
  Using line 1: $\ln 9 = 1\cdot6 \ln2 + c$
  $\Rightarrow c = \ln 9 - 1\cdot6 \ln2 = 1\cdot09$ (3 s.f.)
  And the relationship: $\ln y = 1\cdot60 \ln x + 1\cdot09$

  (See later about modelling situations using the log and exponential functions.)
  At that stage you will see that mathematicians like to simplify this relationship until they have one of the form $y = f(x)$.
  Strengthen your skills with logs by following this argument:
  $\ln y = 1\cdot60 \ln x + 1\cdot09$
  $\Rightarrow \ln y - 1\cdot60 \ln x = 1\cdot09$
  $\Rightarrow \ln y - \ln x^{1\cdot60} = 1\cdot09$
  $\Rightarrow \ln(y/ x^{1\cdot60}) = 1\cdot09$
  $\Rightarrow y/x^{1\cdot60} = e^{1\cdot09} = 3$
  $\Rightarrow y = 3x^{1\cdot60})$

- Solve for a and b given two pairs of corresponding values of x and y:

  Type 2: $y = ax^b$

  This again is most likely to occur in a situation where you are given a graph, told that the equation of the curve is of the form $y = ax^b$ and given/shown two points the curve passes through.

Example

A curve has an equation of the form $y = ax^b$ where a and b are constants.

The curve passes through the points (2, 1·6) and (4, 12·8).

Find the equation of the curve.

Using clue 1: $1·6 = a \times 2^b$

Using clue 2: $12·8 = a \times 4^b$

The trick is to divide ... and the a's will *cancel*:

$$\frac{12·8}{1·6} = a \times 4^b = \frac{4^b}{2^b} = \left(\frac{4}{2}\right)^b$$

$\Rightarrow 8 = 2^b$

$\Rightarrow \ln(8) = \ln(2^b)$

$\Rightarrow \ln(8) = b\ln(2)$

$\Rightarrow b = \ln(8)/\ln(2) = 3$

In this particular example you could have spotted the answer earlier, but it is the general technique that is being shown.

Now return to clue 1:

$1·6 = a \times 2^3$

$\Rightarrow a = 1·6/8 = 0·2$.

The equation of the curve is $y = 0·2x^3$.

Definitely an A/B problem.

- Solve for a and b given two pairs of corresponding values of x and y:

  Type 3

  $y = ab^x$

  Again this type of problem usually comes from a graphical context.

Example

A curve has an equation of the form $y = ab^x$ where a and b are positive constants.

The curve passes through the points (1, 0·7) and (3, 1·2).

Find the equation of the curve.

(Response) ──

Using clue 1: $0·7 = a \times b^1$.

Using clue 2: $1·2 = a \times b^3$.

Again the trick is to divide ... and the a's will *cancel*:

$\Rightarrow {}^{1·2}\!/_{0·7} = b^2$

$\Rightarrow b = 1·3$.

Using clue 1 again:

$0·7 = a \times 1·3^1$

$\Rightarrow a = 0·7/1·3 = 0·54$.

The equation of the curve is $y = 0.54 \times 1.3^x$.

This question is also designated as a type A/B question. —————⊂⊃

- Use a straight line graph to confirm a relationship of the form
  $y = ax^b$, also $y = ab^x$ [a A/B level skill]

Type 1

If x and y are related so that $y = ax^b$,
taking the log of both sides ...

$\log(y) = \log(ax^b)$

$\Rightarrow \log(y) = \log(a) + \log(x^b)$

$\Rightarrow \log(y) = b\log(x) + \log(a)$

Remember a, and hence $\log(a)$, and b are just constants.
If we draw a graph plotting $\log x$ against $\log y$ we will get a straight line
cutting the $\log(y)$ axis at $\log(a)$ and having a gradient of b.
Seeing that straight line confirms the relationship.

Example
Consider the relation $y = 3.2x^{1.4}$.
Here is a table of some values
and an accompanying graph:

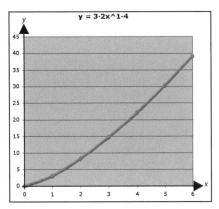

y = 3·2x^1·4

x	0	1	2	3	4	5	6
y	0	3·2	8·44	14·9	22·3	30·5	39·3

Note that the curve passes through the
origin.
Now we take the logs of both x and y ...

log(x)	–	0	0·30	0·48	0·61	0·70	0·78
log(y)	–	0·51	0·93	1·178	1·35	1·48	1·59

... ignoring the (0, 0) ... the log(0) is
undefined.

... And the points produce a straight line
graph.

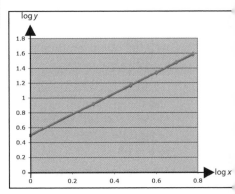

Note that the 'y-intercept' is $0·51$ ...$0·51 = \log a \Rightarrow a = 10^{0·51} = 3·2$
and the gradient is $(0·93 - 0·51)/(0·30 - 0) = 1·4 \Rightarrow b = 1·4$
So x and y are related by the formula: $y = 3·2x^{1·4}$.

Type 2
If x and y are related so that $y = ab^x$,
taking the log of both sides ...
$\log(y) = \log(ab^x)$
$\Rightarrow \log(y) = \log(a) + \log(b^x)$
$\Rightarrow \log(y) = x\log(b) + \log(a)$

Remember a and b, and hence $\log(a)$ and $\log(b)$ are just constants.
If we draw a graph plotting x against $\log y$ we will get a straight line
cutting the $\log(y)$ axis at $\log(a)$ and having a gradient of $\log(b)$.
Seeing that straight line confirms the relationship.

Example
Consider the relation $y = a.b^x$
Here is a table of some values and an
accompanying graph.

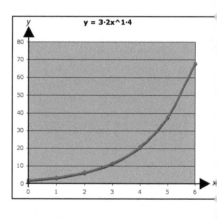

x	0	1	2	3	4	5	6
y	2	3·6	6·48	11·6	21·0	37·8	68·0

Note that the curve does not pass through
the origin.
Now we map x against $\log y$ ...

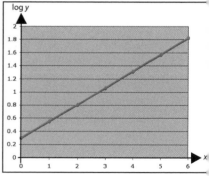

x	0	1	2	3	4	5	6
$\log(y)$	0·301	0·556	0·812	1·07	1·32	1·58	1·83

And the points produce a straight line graph.

Note that the 'y-intercept' is $0·301$
...$0·301 = \log a \Rightarrow a = 10^{0·301} = 2$
and the gradient is
$(0·556 - 0·301)/(1 - 0) = 0·255$
$\Rightarrow \log(b) = 0·255$
$\Rightarrow b = 1·8$

The relationship between x and y is : $y = 2 \times 1·8^x$

- Model mathematically situations involving the logarithmic or exponential function.

  If I were a scientist and my data produced a simple curve passing through the origin, I would explore the possibility that my data could be modelled by $y = ax^b$ and proceed to find the constants of the model by the method described in type 1.

  If my data produced a simple curve which did not pass through the origin I would explore the possibility that the model was $y = ab^x$ and proceed to find a and b by the method shown in type 2.

  Occasionally in the exam you are asked to justify why getting a straight line graph when you plot x against log y (or log x against log y) should indicate a model of the form $y = ax^b$ (or $y = ab^x$). These are grade A marks ... you should familiarize yourself with the argument:

  $$y = ax^b$$
  $\Leftrightarrow \quad \log(y) = \log(ax^b)$
  $\Leftrightarrow \quad \log(y) = \log(a) + \log(x^b)$
  $\Leftrightarrow \quad \log(y) = b\log(x) + \log(a)$

  which is of the form $Y = mX + c$ ... a straight line ... and be able to run it backwards or forwards.

  Several years ago the question appeared in the exam and was greeted dramatically by the press as 'The Killer Question'. It is grade A/B but it's not as difficult as the press make out.

  The SQA provide some notes in the conditions and arrangements that are worthy of mention:

  (i) Change of base is not included ... it can be a handy thing to know though.

  Briefly it states $\log_b a = \dfrac{\log a}{\log b}$ in any base you fancy!

  So, for example $\log_5 2 = \dfrac{\log_{10} 2}{\log_{10} 5} = 0.431$

  ... or even ... $\log_5 2 = \dfrac{\ln 2}{\ln 5} = 0.431$

  (ii) When base is understood in a particular context, log x can be used for $\log_a x$.

  The base a will normally be 10 or e.

  (iii) The notation ln x should be introduced to candidates.

  (iv) Numerical uses of logarithms and exponential evaluations are best done by calculators ... and if your calculator doesn't have a 'log to any base' button, the change of base formula could come in handy.

Objective questions

1 What is $\log(a^2b)$ equal to?

   A $2\log a + \log b$
   B $2(\log a + \log b)$
   C $2 \log a \log b$
   D $\log a + \log 2 + \log b$

Rough working

$\log(a^2b) = \log(a^2) + \log(b) = 2\log(a) + \log(b)$

Choose option A

A little tip when you are simplifying logs ... don't look at the distractors until you have your answer.

2 What is the value of $\log_3\left(\dfrac{1}{9}\right)$?

   A   2
   B   −2
   C   3
   D   −3

Rough working

$\log_3\left(\dfrac{1}{9}\right) = \log_3(3^{-2}) = -2\log_3(3) = -2$

Choose option B

3 $\log_a 16 = \frac{1}{2}$. What is the value of a?

   A   8
   B   4
   C   256
   D   32

Rough working

$\log_a 16 = \frac{1}{2} \Leftrightarrow 16 = a^{\frac{1}{2}} \Rightarrow a = 16^2 = 256$

Choose option C

4 Given that ln $xy = 0$. What does that imply about x and y?

   A   ln x and ln $y = 0$
   B $x + y = 0$
   C $x + y = 1$
   D $xy = 1$

Rough working

$\ln xy = 0 \Rightarrow xy = e^0 = 1$

Choose option D

⑤ Which of the following could be a sketch of $y = \log_5 x$?

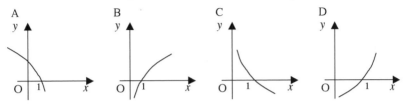

A   B   C   D

Rough working

We're looking for an increasing function ... ruling out A and C.
We're looking for a function with a decreasing gradient ... ruling out D.

Choose option B

⑥ $e^x > e^{2x}$. What does that imply?

A $x > 0$
B $x < 0$
C $e^x > 1$
D   Nothing. It's impossible.

Rough working

$f(x) = e^x$ is an increasing function so $f(x_a) > f(x_b) \Rightarrow x_a > x_b$
So $e^x > e^{2x} \Rightarrow x > 2x \Rightarrow 0 > x$

Choose option B

⑦ $3 \times 8^x = 6$. What is the value of x?

A -3
B $\frac{1}{2}$
C $\frac{1}{3}$
D $\frac{1}{4}$.

Rough working

$3 \times 8^x = 6$
$\Rightarrow 8^x = 2$
$\Rightarrow (2^3)^x = 2^1$
$\Rightarrow 2^{3x} = 2^1$
$\Rightarrow 3x = 1$
$\Rightarrow x = \frac{1}{3}$

Choose option C

❽ log *y* was graphed against log *x* and a straight line was produced.
This shows that $y = ax^b$ where *a* and *b* are constants.
What is the value of *b*?

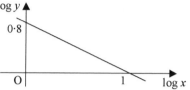

A 0·8
B −0·8
C log 0·8
D ¾

Rough working

$y = ax^b$
$\Rightarrow \log y = \log a + b\log x$
b is the gradient of the line.
gradient = − 0·8

Choose option B

A short response question that doesn't need a calculator

❾ 3log 2 + ½ log 9 − log 6 = log *k*.
Find *k*.

Response

3log 2 + ½ log 9 − log 6 = log *k*.
$\Rightarrow \log 2^3 + \log 9^{½} - \log 6 = \log k.$
$\Rightarrow \log 8 + \log 3 - \log 6 = \log k.$
$\Rightarrow \log (8 \times 3 \div 6) = \log k.$
$\Rightarrow \log (4) = \log k.$
$\Rightarrow k = 4$

Marking scheme

- •¹ use law of logs to deal with coefficients   *ss*
- •² use law of logs to combine logs   *ss*
- •³ processing   *pd*
- •⁴ knowing log *a* = log *b* $\Rightarrow$ *a* = *b*   *ic*

A short response question that needs a calculator

⑩ Given that $x = \log_3 5 + \log_3 6$ find *algebraically* the value of x. *4 marks*

The 'codeword' *algebraically* is telling you that answers based on trial and error, the use of graphical calculators or the use of special functions on a calculator will receive no marks. This question is asking you to show your dexterity in the use of the laws of logs. The calculator should only be used at the end to evaluate a log or logs to the base 10 or e. This compares with the Standard Grade question which asks you to calculate the standard deviation. No marks are awarded for simply entering data and pushing the 'standard deviation' button. It is the technique that is being tested.

(Response) ─────────────────────────────────

$x = \log_3 5 + \log_3 6$

$\Rightarrow x = \log_3 30$

$\Rightarrow 3^x = 30$ ... [by the definition of a log]

$\Rightarrow \log(3^x) = \log 30$ ... using either base 10 or e.

$\Rightarrow x \log 3 = \log 30$

$\Rightarrow x = \frac{\log 30}{\log 3}$

$\Rightarrow x = 3 \cdot 10$ (3 s.f.) ... using the calculator. ────────⊂────⊃

Marking scheme

●¹	use law of logs	*ss*
●²	use definition to write in index form	*ss*
●³	take logs of both sides	*ss*
●⁴	process $\frac{\log 30}{\log 3}$	*pd*

Note that although the changing of base is not formally required, this is effectively what happens here; the base is changed from 3 to 10 to allow the use of standard tables/calculator.

Here are a couple of SQA questions showing some of the modelling features.
First, 'The Killer Question' (from 1998 paper II Q 11)

(a) **The variables x and y are connected by a relationship of the form $y = ae^{bx}$ where a and b are constants. Show that there is a linear relationship between $\log_e y$ and x.**

3 marks

(b) **From an experiment some data were obtained. The table shows the data which lies on the line of best fit.**

6 marks

x	3·1	3·5	4·1	5·2
y	21 876	72 631	439 392	11 913 076

The variables x and y in the above table are connected by a relationship of the form $y = ae^{bx}$. Determine the values of a and b.

Response

(a) $y = ae^{bx}$ ... take the natural log of both sides (i.e. log to base e)

$\Rightarrow \log_e(y) = \log_e(ae^{bx})$

$\Rightarrow \log_e(y) = \log_e(a) + \log_e(e^{bx})$

$\Rightarrow \log_e(y) = \log_e(a) + bx \log_e(e)$ *These steps are a standard piece of theory.*

$\Rightarrow \log_e(y) = \log_e(a) + bx$

$\Rightarrow \log_e(y) = bx + \log_e(a)$

Compare this with the linear function $Y = mX + c$.
If you plot $\log_e(y)$ against x you'll get a straight line of gradient b and y-intercept $\log_e(a)$.

(b) Remember we plotted $\log_e y$ against x ... rewrite the table to show what we plotted.

x	3·1	3·5	4·1	5·2
$\log_e y$	9·99	11·2	13·0	16·3

Next we want the gradient and y-intercept ... pick any two points to get the gradient say (3·1, 9·99) and (4·1, 13·0)

... $b = \dfrac{13\cdot0 - 9\cdot99}{4\cdot1 - 3\cdot1} = 3\cdot0$ (to 2 s.f.)

... use $b = 3$ given that the data is experimental anyway.

So the equation is of the form $\log_e(y) = 3x + \log_e(a)$

Plug (4·1, 13) into this to find a ...

$13\cdot0 = 3 \times 4\cdot1 + \log_e(a)$

$\Rightarrow \log_e(a) = 0\cdot7$

$\Rightarrow a = e^{0\cdot7} = 2\cdot0$ (to 2 s.f.)

So the relationship is $y = 2e^{3x}$. ⟵─────────────────⊂⊃

Marking scheme

- ●¹ take natural logs of both sides   ss
- ●² apply the laws of logs   pd
- ●³ communicate the linear form   ic
- ●⁴ interpret straight line: gradient as b   ic
- ●⁵ take logs of at least two values   ss
- ●⁶ evaluate gradient   pd
- ●⁷ substitute to find intercept   ss
- ●⁸ strategy to solve $\log_e(a) = 0\cdot7$ ss
- ●⁹ process the data to get a   pd

This is an A/B grade question with early chances for the C-grade candidate to pick up a few marks. The early part of the question is standard.

In the year 2000 another example of this area of work appeared.

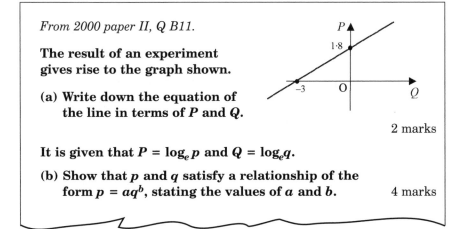

From 2000 paper II, Q B11.

The result of an experiment gives rise to the graph shown.

(a) Write down the equation of the line in terms of P and Q.

2 marks

It is given that $P = \log_e p$ and $Q = \log_e q$.

(b) Show that p and q satisfy a relationship of the form $p = aq^b$, stating the values of a and b.

4 marks

(Response)

(a) The equation is of the form $y = mx + c$.
   By inspection $c = 1.8$
   $m = {}^{1.8}\!/_3 = 0.6$
   $x = Q$ and $y = P$
   Equation of line: $P = 0.6Q + 1.8$.

(b) Substituting $P = \log_e p$ and $Q = \log_e q$ we get:
   $\log_e p = 0.6 \log_e q + 1.8$
   $\Rightarrow \log_e p = 0.6 \log_e q + \log_e e^{1.8}$
   $\Rightarrow \log_e p = \log_e q^{0.6} + \log_e e^{1.8}$
   $\Rightarrow \log_e p = \log_e(q^{0.6} e^{1.8})$
   $\Rightarrow p = q^{0.6} e^{1.8}$
   $\Rightarrow p = 6.0 q^{0.6}$
   So $a = 6$ and $b = 0.6$

Marking scheme

- $\bullet^1$ interpret gradient   *ic*
- $\bullet^2$ state equation of line   *ic*
- $\bullet^3$ substitute into the linear form   *ic*
- $\bullet^4$ laws of logs for coefficient   *ss*
- $\bullet^5$ laws of logs to simplify   *ss*
- $\bullet^6$ arrange to final form and identify a and b *ic*

In this example working in base e is critical ... so working in base 10 will lose a mark.

These questions tend to come later on in a paper. They are typically A/B grade questions with a C grade entry. In this case the C grade entry was simple straight-line work.

You should get familiar with both arguments:
(i)   If a relation of the form $y = ax^b$ exists then plotting **log y** against **log x** will produce a straight line with gradient b and y-intercept log a.
(ii)   If a relation of the form $y = ab^x$ exists then plotting **log y** against x will produce a straight line with gradient log b and y-intercept log a.

Only when the form is $y = ae^{bx}$ are you forced to use e as the base. However this is only a slight variation of type (ii) as you will have seen above.

Here is an example taken from real data:

The table shows the planets, their distance from the sun in convenient units and the time it takes them to go round the sun (their *period*).

Planet	distance (A.U.)	period
Mercury	0.39	0.24
Venus	0.72	0.62
Earth	1	1
Mars	1.52	1.88
Ceres	2.77	5
Jupiter	5.2	12
Saturn	9.54	29
Uranus	19.18	84
Neptune	30.06	165
Pluto	39.52	248

An astronomer drew a graph of the data.

Notice the graph passes through the origin. This sort of graph leads us to believe that the relationship is $P = aD^b$ where a and b are constants, P is the period in years and D is the distance.
If we graph log D against log P we get a straight line.

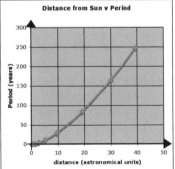

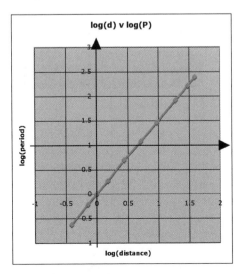

log(distance)	log(period)
−0.409	−0.620
−0.143	−0.208
0	0
0.182	0.274
0.442	0.699
0.716	1.08
0.980	1.46
1.28	1.92
1.48	2.22
1.60	2.39

How do we find a and b and hence the relationship between *period* and *distance*?

Remember the theory:
$P = aD^b$
$\Rightarrow \log P = \log(aD^b)$
$\Rightarrow \log P = \log(a) + b \log(D)$.

Graphing $\log P$ against $\log D$ gives ... a straight line cutting the y axis at $(0, \log(a))$ with a gradient of b.
From the table $(0, 0)$ is on the line. $\Rightarrow \log a = 0 \Rightarrow a = 1$.
Using any two points on the line, find the gradient.
Use $(0, 0)$ and $(0\cdot182, 0\cdot274)$
$b = (0\cdot274 - 0)/(0\cdot182 - 0) = 1\cdot5$

The relationship is $P = D^{1\cdot5}$.

This is the actual relationship discovered by Kepler in 1619.

How might a question like this be exploited by the SQA?
It is obviously impractical to give you the table of distances and periods and expect you to reproduce all the above steps.
It is possible that:
(i)   they give you a graph and ask you to justify why the model is $P = aD^b$, asking you to go through the steps to linearisation;
(ii) they could give the linear graph of $\log D$ against $\log P$ and ask you to justify the model;
(iii) they could give you the raw data, the linear graph and ask you to find the constants.

These would all be grade A/B questions.

An extended example question that needs a calculator

To find how deep a well is, you can drop a stone in it and see how long it takes to fall. The table shows some data.

Time of fall (t seconds)	Depth of well (d metres)
1	5
2	20
3	45
4	80
5	125
6	180
7	245

Graphing this data has led us to believe that t and d are related by the formula $d = at^b$ where a and b are constants.

(a) Show that the relation between log t and log d is linear. *3 marks*

(b) Find the constants a and b and hence the relationship between t and d. *5 marks*

Response

(a) $d = at^b$

$\Rightarrow \log d = \log(at^b)$

$\Rightarrow \log d = \log(a) + \log(t^b)$

$\Rightarrow \log d = \log(a) + b\log(t)$

Graphing log d against log t will give us a straight line with y-intercept log a and gradient b.

(b) Take the logs of a few values ... here the data for 2 seconds and for 7 seconds has been used.

log (time)	log (depth)
0·3	1·3
0·85	2·39

Find the gradient: $b = (2·39 - 1·3)/(0·85 - 0·3) = 2·0$.

Use the first point to find a: $\log d = \log(a) + b\log(t)$

$\Rightarrow 1·3 = \log(a) + 2 \times 0·3$

$\Rightarrow \log(a) = 1·3 - 2 \times 0·3$

$\Rightarrow \log(a) = 1·3 - 2 \times 0·3 = 0·7$

$\Rightarrow a = 10^{0·7}$

$\Rightarrow a = 5$

The required relationship is: $d = 5t^2$.

Marking scheme

•^{1}	take logs of both sides	ss
•^{2}	use laws of logs	pd
•^{3}	identify equation of line	ic
•^{4}	take logs of two 'points'	ss
•^{5}	find gradient	pd
•^{6}	substitute to find log a	ss
•^{7}	evaluate a and b	pd
•^{8}	declare model	ic

In this case we could have made the finding of a simpler because the point (1, 5) is part of the original data (t, d).

So (0, 0·7) lies on the (log t, log d) graph ... i.e. the y-intercept is 0·7.

However we wished to show a general method for finding a ... you can't always be sure of getting such a convenient point.

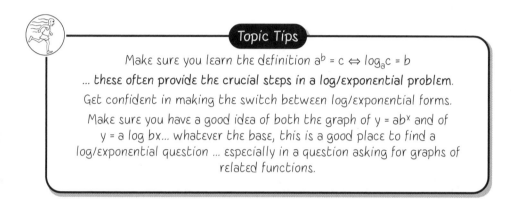

Topic Tips

Make sure you learn the definition $a^b = c \Leftrightarrow \log_a c = b$

... these often provide the crucial steps in a log/exponential problem.

Get confident in making the switch between log/exponential forms.

Make sure you have a good idea of both the graph of $y = ab^x$ and of $y = a \log bx$... whatever the base, this is a good place to find a log/exponential question ... especially in a question asking for graphs of related functions.

15 Further Trigonometric Relationships

What you should know

You are expected to know the following facts.

- How to express $a \cos \theta + b \sin \theta$ in the form $r \cos(\theta \pm \alpha)$ or $r \sin(\theta \pm \alpha)$.
  There is usually an assumption that $r > 0$ and that $0° \le \alpha° \le 360°$ or $0 \le \alpha \le 2\pi$.
  These restrictions may turn up in questions in the exam but will be generally assumed to be the case in the examples below.

To perform such a task we expand the form we wish to achieve and equate the coefficients with a and b.
The task should be able to be done using degrees or radians.

> $\sin(A + B) = \sin A \cos B + \cos A \sin B$

Example 1
Express $3\cos x° + 4 \sin x°$ in the form $r \sin (x + \alpha)°$

Step 1: Expand the form we want ... $r \sin (x + \alpha)° = r \sin x° \cos \alpha° + r \cos x° \sin \alpha°$

Step 2: Compare what we want with what we've got

$3 \cos x° + \underline{4} \sin x° = \underline{r} \sin x° \underline{\cos \alpha°} + r \cos x° \; \textbf{sin } \alpha°$

Equate the coefficients of cos x: **3 = r sin α°**

<u>Equate the coefficients of sin x: 4 = r cos α°</u>

Step 3: Square and add the equations: $3^2 + 4^2 = r^2 \sin^2\alpha° + r^2 \cos^2\alpha°$

$\Rightarrow 25 = r^2 (\sin^2\alpha° + \cos^2\alpha°)$

$\Rightarrow 25 = r^2$ ... since $(\sin^2\alpha° + \cos^2\alpha°) = 1$ (Standard Grade)

$\Rightarrow r = 5$

Step 4: Divide the equations, the sin term by the cos term.

$\dfrac{r \sin \alpha°}{r \cos \alpha°} = \dfrac{3}{4} \Rightarrow \tan \alpha° = 0{\cdot}75 \Rightarrow \alpha = 36{\cdot}9$ or $180 + 36{\cdot}9 = 216{\cdot}9$.

However, from the original equations we see that sin α is positive (1st or 2nd quadrant) and that cos α is positive (1st or 4th quadrant) ... so it is a 1st quadrant angle.

So α = 36·9.

Step 5: Answer the question ... $3 \cos x° + 4 \sin x° = 5 \sin (x + 36{\cdot}9)°$

$\cos(A + B) = \cos A \cos B - \sin A \sin B$

Example 2

Express $3 \cos x° + 4 \sin x°$ in the form $r \cos (x + \alpha)°$

Step 1: Expand the form we want ... $r \cos (x + \alpha)° = r \cos x° \cos \alpha° - r \sin x° \sin \alpha°$

Step 2: Compare what we want with what we've got:

$3 \cos x° \underline{+ 4} \sin x° = \underline{r} \cos x° \; \textbf{cos } \alpha° \underline{- r} \sin x° \underline{\sin \alpha°}$

Equate the coefficients of cos x°:   3 = r cos α°

<u>Equate the coefficients of sin x°:   4 = − r sin α° ⇒ r sin α° = − 4</u>

Step 3: Square and add the equations: $3^2 + 4^2 = r^2 \cos^2\alpha° + r^2 \sin^2\alpha°$

$\Rightarrow 25 = r^2 (\cos^2\alpha° + \sin^2\alpha°)$

$\Rightarrow 25 = r^2$

$\Rightarrow r = 5$

Step 4: Divide the equations, the sin term by the cos term.

$\dfrac{r \sin \alpha°}{r \cos \alpha°} = \dfrac{-4}{3} \Rightarrow \tan \alpha° = -1{\cdot}333$

$\Rightarrow \alpha = -53{\cdot}1$ or $180 + (-53{\cdot}1) = 126{\cdot}9$ or $180 + 126{\cdot}9 = 306{\cdot}9$

However, from the original equations we see that sin α is negative (3rd or 4th quadrant) and that cos α is positive (1st or 4th quadrant) ... so it is a 4th quadrant angle.

So α = 306·9.

Step 5: Answer the question ... $3 \cos x° + 4 \sin x° = 5 \cos (x + 306{\cdot}9)°$

Each of these steps should be obvious in any response – communication and strategy marks can go missing if they aren't.

The deliberation as to what quadrant α is in can be shortened to a diagram.

e.g.

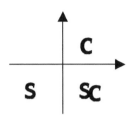

The Cs indicating the cos possibilities;
the Ss representing the sin possibilities;
the quadrant with two symbols is the one we want

You will not always be given guidance as to which of the four possible forms is wanted. Often, when problem solving, the particular form is unimportant. In these cases you have license to pick whichever form you fancy.

We recommend using $r\sin(\theta \pm \alpha)$, picking the sign to 'agree' with that in the given form. There is less chance of error and α is in the 1st quadrant.

This is a regular question in the exam. Make sure your communication skills are good and make all five steps evident.

As an aid to understanding this, an interactive spreadsheet can be very beneficial. Instructions on how to make such a spreadsheet are included in Appendix 5 on *www.leckieandleckie.co.uk*.

The same type of question can be asked using radians as the unit of measure. In these cases you are generally safer to work in radians throughout. It is a bad habit, and dangerous, to switch to degrees and then re-establish radian measure at the end.

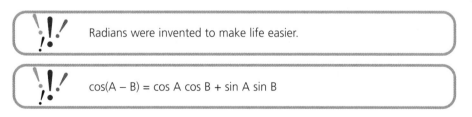

Radians were invented to make life easier.

$\cos(A - B) = \cos A \cos B + \sin A \sin B$

Example

Express $\sqrt{3}\cos x + \sin x$ in the form $k\cos(x - \alpha)$, $0 \leq \alpha \leq 2\pi$

Step 1: Expand the form we want ... $r \cos(x - \alpha) = r \cos x \cos \alpha + r \sin x \sin \alpha$

Step 2: Compare what we want with what we've got.

$\sqrt{3} \cos x \underline{+1} \sin x = r \cos x \, \textbf{cos } \alpha \underline{+ r} \sin x \, \underline{\sin \alpha}$

Equate the coefficients of cos x: $\sqrt{3} = r \cos \alpha$

Equate the coefficients of sin x: $1 = r \sin \alpha$

Step 3: Square and add the equations: $1^2 + (\sqrt{3})^2 = r^2 \cos^2\alpha + r^2 \sin^2\alpha$

$\Rightarrow 4 = r^2 (\cos^2\alpha + \sin^2\alpha)$

$\Rightarrow 4 = r^2$

$\Rightarrow r = 2$

Step 4: Divide the equations, the sin term by the cos term.

$\dfrac{r \sin \alpha}{r \cos \alpha} = \dfrac{1}{\sqrt{3}} \Rightarrow \tan \alpha = \tfrac{1}{\sqrt{3}}$

$\Rightarrow \alpha = \tfrac{\pi}{6} \text{ or } \pi + \tfrac{\pi}{6} = \tfrac{7\pi}{6}$ ...(Think of π as $\tfrac{6\pi}{6}$)

However, from the original equations we see that sin α is positive
(1st or 2nd quadrant) and that cos α is positive (1st or 4th quadrant) ... so it is a
1st quadrant angle.

So $\alpha = \tfrac{\pi}{6}$.

Step 5:   Answer the question ... $\sqrt{3} \cos x + \sin x = 2 \cos(x - \tfrac{\pi}{6})$.

Related problems

You should be able to use the above knowledge to solve the following kinds of
problems.

- Express $a \cos \theta + b \sin \theta$ in the form $r \cos(\theta \pm \alpha)$ or $r \sin(\theta \pm \alpha)$.
  This has been explored above but for the sake of completeness we'll explore
  another form in an example.

Example

Express $\sin x° - 2 \cos x°$ in the form $r \cos(x + \alpha)°$

Step 1: Expand the form we want. $r \cos(x + \alpha)° = r \cos x° \cos \alpha° - r \sin x° \sin \alpha°$

Step 2: Compare what we want with what we've got.

$\underline{1}\sin x° - \textbf{2} \cos x° = r \cos x° \, \textbf{cos } \alpha° \, \underline{- r} \sin x° \, \underline{\sin \alpha°}$

Equate the coefficients of cos x°: $-2 = r \cos \alpha°$

Equate the coefficients of sin x°: $1 = -r \sin \alpha° \Rightarrow r \sin \alpha° = -1$

Step 3: Square and add the equations: $1^2 + (-2)^2 = r^2 \cos^2\alpha° + r^2 \sin^2\alpha°$

$\Rightarrow 5 = r^2 (\cos^2\alpha° + \sin^2\alpha°)$

$\Rightarrow 5 = r^2$

$\Rightarrow r = \sqrt{5}$

Step 4: Divide the equations, the sin term by the cos term.

$\dfrac{r \sin \alpha°}{r \cos \alpha°} = \dfrac{-1}{-2} \Rightarrow \tan \alpha° = 26 \cdot 6$

$\Rightarrow \alpha = 26 \cdot 6 \text{ or } 180 + 26 \cdot 6 = 206 \cdot 6$

However, from the original equations we see that sin $\alpha°$ is negative (3rd or 4th quadrant) and that cos $\alpha°$ is negative (2nd or 3rd quadrant) ... so it is a 3rd quadrant angle.

So $\alpha = 206·6$.

Step 5: Answer the question ... sin $x°$ − 2 cos $x°$ = $\sqrt{5}$ cos $(x + 206·6)°$.

- Solve equations of the form $a \cos \theta + b \sin \theta = c$

Example

Solve 2 sin $x°$ − 3 cos $x°$ = 2, $0 \le x \le 360$

First we express the left-hand side in a suitable form.

No one has dictated the form so use whichever you like, say, $r \sin(x - \alpha)°$.

> sin$(A + B)$ = sin A cos B − cos A sin B

Step 1: $r \sin (x - \alpha)° = r \sin x° \cos \alpha° - r \cos x° \sin \alpha°$

Step 2: $2\sin x° - 3 \cos x° = r \sin x° \cos \alpha° - r \cos x° \sin \alpha°$

Equate the coefficients of cos $x°$: $-3 = -r \sin \alpha° \Rightarrow r \sin \alpha° = 3$

Equate the coefficients of sin $x°$: $2 = r \cos \alpha°$

Step 3: Square and add the equations: $2^2 + 3^2 = r^2 \cos^2\alpha° + r^2 \sin^2\alpha°$

$\Rightarrow 13 = r^2 (\cos^2\alpha° + \sin^2\alpha°)$

$\Rightarrow 13 = r^2$

$\Rightarrow r = \sqrt{13}$

Step 4: Divide the equations, the sin term by the cos term.

$\dfrac{r\sin \alpha°}{r\cos \alpha°} = \dfrac{3}{2} \Rightarrow \tan \alpha° = 1·5$

$\Rightarrow \alpha = 56·3$ or $180 + 56·3 = 236·3$

However, from the original equations we see that sin $\alpha°$ is positive (1st or 2nd quadrant) and that cos $\alpha°$ is positive (1st or 4th quadrant) ... so it is a 1st quadrant angle.

So $\alpha = 56·3$

$\Rightarrow 2 \sin x° - 3 \cos x° = \sqrt{13} \sin (x - 56·3)°$

$2 \sin x° - 3 \cos x° = 2$

$\Rightarrow \sqrt{13} \sin (x - 56·3)° = 2$

$\Rightarrow \sin (x - 56·3)° = 2/\sqrt{13} = 0·555$

$\Rightarrow x - 56·3 = 33·7, 180 - 33·7 = 146·3$

$\Rightarrow x = 90, 202·6$

Again you could be asked to handle radians.

Example

Solve $\sin x - \cos x = 1$, $0 \leq x \leq 2\pi$

You haven't been told the form wanted so pick your own, say, $r\sin(x - \alpha)$.

Step 1: Expand the form we want … $r\sin(x - \alpha) = r\sin x \cos\alpha - r\cos x \sin\alpha$

Step 2: Compare what we want with what we've got

$\underline{1}\sin x - \mathbf{1}\cos x = \underline{r}\sin x \underline{\cos\alpha} - r\cos x \,\mathbf{\sin\alpha}$

Equate the coefficients of cos x: $-1 = -r\sin\alpha \Rightarrow \mathbf{1 = r\sin\alpha}$

<u>Equate the coefficients of sin x:</u> $1 = r\cos\alpha$

Step 3: Square and add the equations: $(1)^2 + (1)^2 = r^2\cos^2\alpha + r^2\sin^2\alpha$

$\Rightarrow 2 = r^2(\cos^2\alpha + \sin^2\alpha)$

$\Rightarrow 2 = r^2$

$\Rightarrow r = \sqrt{2}$

Step 4: Divide the equations, the sin term by the cos term.

$\dfrac{r\sin\alpha}{r\cos\alpha} = \dfrac{1}{1} = 1 \Rightarrow \tan\alpha = 1$

$\Rightarrow \alpha = \frac{\pi}{4}$ or $\pi + \frac{\pi}{4} = \frac{5\pi}{4}$ …(Think of π as $\frac{4\pi}{4}$)

However, from the original equations we see that $\sin\alpha$ is positive (1st or 2nd quadrant) and that $\cos\alpha$ is positive (1st or 4th quadrant) … so it is a 1st quadrant angle.

So $\alpha = \frac{\pi}{4}$.

Step 5: $\sin x - \cos x = \sqrt{2}\sin(x - \frac{\pi}{4})$

Step 6: Now answer the question.

$\sin x - \cos x = 1$

$\Rightarrow \sqrt{2}\sin(x - \frac{\pi}{4}) = 1$

$\Rightarrow \sin(x - \frac{\pi}{4}) = \frac{1}{\sqrt{2}}$

$\Rightarrow x - \frac{\pi}{4} = \frac{\pi}{4}, \pi - \frac{\pi}{4} = \frac{3\pi}{4}, \dots$

$\Rightarrow x = \frac{\pi}{2}, \pi$

In this type of question, marks will be deducted if you persist in working in degrees.

- Find maximum and minimum values of expressions of the form $a\cos\theta + b\sin\theta$

Example

Find the maximum and minimum values of $\sin x° - 3\cos x°$.

You can pick the form you fancy.

Just for the sake of it let's pick $r\cos(x + \alpha)°$.

Step 1: $r\cos(x + \alpha)° = r\cos x° \cos\alpha° - r\sin x° \sin\alpha°$

Step 2: $\sin x° - 3\cos x° = r\cos x° \cos\alpha° - r\sin x° \sin\alpha°$

Equate the coefficients of cos $x°$: $-3 = r\cos\alpha° \Rightarrow r\cos\alpha° = -3$

Equate the coefficients of sin $x°$: $1 = -r\sin\alpha° \Rightarrow r\sin\alpha° = -1$

Step 3: Square and add the equations: $1^2 + (-3)^2 = r^2 \cos^2\alpha° + r^2 \sin^2\alpha°$

$$\Rightarrow 10 = r^2 (\cos^2\alpha° + \sin^2\alpha°)$$
$$\Rightarrow 10 = r^2$$
$$\Rightarrow r = \sqrt{10}$$

Step 4: Divide the equations, the sin term by the cos term.

$$\frac{r\sin \alpha°}{r\cos \alpha°} = \frac{-1}{-3} \Rightarrow \tan \alpha° = \tfrac{1}{3}$$

$$\Rightarrow \alpha = 18{\cdot}4 \text{ or } 180 + 18{\cdot}4 = 198{\cdot}4$$

However, from the original equations we see that $\sin \alpha°$ is negative (3rd or 4th quadrant) and that $\cos \alpha°$ is negative (2nd or 3rd quadrant) ... so it is a 3rd quadrant angle.

So $\alpha = 198{\cdot}4$

$$\sin x° - 3 \cos x° = \sqrt{10} \cos (x + 198{\cdot}4)°$$

Now answer the question.

$$-1 \le \cos (x + 198{\cdot}4)° \le 1$$
$$\Rightarrow -\sqrt{10} \le \sqrt{10} \cos (x + 198{\cdot}4)° \le \sqrt{10}$$

Maximum value $\sqrt{10}$; minimum value $-\sqrt{10}$

Again radian measure may be asked for.

Example

Find the maximum and minimum value of $\sin x - \sqrt{3} \cos x + 1$.

Note that the $+ 1$ is just an added bit to the problem ... first let's find a convenient form for $\sin x - \sqrt{3} \cos x$ ... say $r \sin(x - \alpha)$.

Expand: $\sin x - \sqrt{3} \cos x = r \sin x \cos \alpha - r \cos x \sin \alpha$

Equate coefficients: $-\sqrt{3} = - r \sin \alpha;\ 1 = r \cos \alpha$

Square and add: $r^2 \sin^2 x + r^2 \cos^2 x = (\sqrt{3})^2 + 1^2 = 4$

$$\Rightarrow r^2 (\sin^2 x + \cos^2 x) = 4$$
$$\Rightarrow r = 2$$

Divide: $\dfrac{r\sin \alpha}{r\cos \alpha} = \dfrac{\sqrt{3}}{1}$

$$\Rightarrow \alpha = \tfrac{\pi}{3}$$

Complete: $\Rightarrow \sin x - \sqrt{3} \cos x = 2 \sin(x - \tfrac{\pi}{3})$

$$\Rightarrow \sin x - \sqrt{3} \cos x + 1 = 2 \sin(x - \tfrac{\pi}{3}) + 1$$

Now answer the question:

$$-1 \le \sin(x - \tfrac{\pi}{3}) \le 1$$
$$\Rightarrow -2 \le 2\sin(x - \tfrac{\pi}{3}) \le 2$$
$$\Rightarrow -1 \le 2\sin(x - \tfrac{\pi}{3}) + 1 \le 3$$

Maximum value 3; minimum value -1.

- Find values of θ corresponding to maximum/minimum values of expression.

Example 1

Find the maximum and minimum values of $\sin x° - 3 \cos x°$ ... and say for what values of x between 0 and 360 these values occur.

First we express $\sin x° - 3 \cos x°$ in a suitable form

$\sin x° - 3 \cos x° = \sqrt{10} \cos (x + 198\cdot4)°$ ... see above.

$\Rightarrow \sqrt{10} \le \sqrt{10} \cos (x + 198\cdot4)° = \sqrt{10}$

We know from above that the maximum is $\sqrt{10}$.

This occurs when $\sqrt{10} \cos (x + 198\cdot4)° = \sqrt{10}$

$\Rightarrow \cos (x + 198\cdot4)° = 1$

$\Rightarrow x + 198\cdot4 = 0, 360, ...$

$\Rightarrow x = -198\cdot4, 161\cdot6$

$\Rightarrow x = 161\cdot6$ (given the range)

Similarly, we know from above that the minimum is $-\sqrt{10}$.

This occurs when $\sqrt{10} \cos (x + 198\cdot4)° = -\sqrt{10}$

$\Rightarrow \cos (x + 198\cdot4)° = -1$

$\Rightarrow x + 198\cdot4 = 180, 540, ...$

$\Rightarrow x = -18\cdot4, 341\cdot6$

$\Rightarrow x = 341\cdot6$ (given the range)

This is designated an A/B grade skill.

Example 2 (a radian example)

Find the maximum and minimum value of $\sin x - \sqrt{3} \cos x + 1$ and say for what values of x between 0 and 2π these values occur.

From the example earlier we know $\sin x - \sqrt{3} \cos x + 1 = 2 \sin(x - \frac{\pi}{3}) + 1$ and that $-1 \le 2\sin(x - \frac{\pi}{3}) + 1 \le 3$.

Thus the maximum is 3 and this will occur when $\sin(x - \frac{\pi}{3}) = 1$

$\Rightarrow x - \frac{\pi}{3} = \frac{\pi}{2}$

$\Rightarrow x = \frac{\pi}{2} + \frac{\pi}{3} = \frac{3\pi}{6} + \frac{2\pi}{6} = \frac{5\pi}{6}$

... and the minimum is -1 and this will occur when $\sin(x - \frac{\pi}{3}) = -1$

$\Rightarrow x - \frac{\pi}{3} = \frac{3\pi}{2}$

$\Rightarrow x = \frac{3\pi}{2} + \frac{\pi}{3} = \frac{9\pi}{6} + \frac{2\pi}{6} = \frac{11\pi}{6}$

Finding the maxima and minima can be linked with other topics.
Remember that if $f'(x) > 0$ then the function is always increasing.
You could be asked:
Given $f'(x) = \sin x - \sqrt{3} \cos x + 3$, prove that $f(x)$ is an increasing function.
Using the same argument as above we get $1 \le 2\sin(x - \frac{\pi}{3}) + 3 \le 5$.
The derivative is never less than zero. Hence result.

There is a type of problem that has a slight twist as you get to the end. You must watch out for this.

Example
(a) Express $3 \sin x° + 20 \cos x°$ in the form $k \sin(x - a)°$.

(b) Solve $3 \sin x° + 20 \cos x° = -0·3$.

(Response)

(a)

Step 1: $k \sin (x - a)° = k \sin x° \cos a° - k \cos x° \sin a°$

Step 2: $3 \sin x° + 20 \cos x° = k \sin x° \cos a° - k \cos x° \sin a°$

Equate the coefficients of $\cos x°$: $\quad 20 = -k \sin a° \Rightarrow k \sin a° = -20$

Equate the coefficients of $\sin x°$: $\quad 3 = k \cos a°$

Step 3: Square and add the equations: $3^2 + 20^2 = r^2 \cos^2 a° + r^2 \sin^2 a°$

$\Rightarrow 409 = r^2 (\cos^2 a° + \sin^2 a°)$

$\Rightarrow 409 = r^2$

$\Rightarrow r = \sqrt{409}$

Step 4: Divide the equations, the sin term by the cos term.

$$\frac{r \sin a°}{r \cos a°} = \frac{-20}{3} \Rightarrow \tan a° = {}^{-20}\!/_3$$

$\Rightarrow a = -81·4$ or $180 + (-81·4) = 98·5$ or $180 + (98·5) = 278·5$

However, from the original equations we see that sin a is negative (3rd or 4th quadrant) and that cos a is positive (1st or 4th quadrant) ... so it is a 4th quadrant angle.

So $a = 278·5$

$3 \sin x° + 20 \cos x° = \sqrt{409} \sin (x - 278·5)°$

(b)

$3 \sin x° + 20 \cos x° = -0·3$

$\Rightarrow \sqrt{409} \sin (x - 278·5)° = -0·3$

$\Rightarrow \sin (x - 278·5)° = -0·0148$

$\Rightarrow x - 278·5 = -360·85, -179·15, -0·85, 180·85,$

$\Rightarrow x = -82·35, 99·35, 277·65, 459·35,..$

$\Rightarrow x = 99·35, 277·65$ (in the given range)

Because we were going to be adding 278·5 to all our answers found in line three, we had to consider negative solutions by subtracting 360 from our 'usual' answers.

This is considered an A/B grade skill and you should always consider the possibility of negative answers being brought up into the required range.

Objective questions

1   Given that $3\sin x + 4\cos x = k\cos(x + a)$, $0 \leq a \leq 360$
Which of the following is true?

A $k\cos a = 3$ and $k\sin a = 4$
B $k\cos a = 4$ and $k\sin a = 3$
C $k\cos a = 4$ and $k\sin a = -3$
D $k\cos a = 3$ and $k\sin a = -4$

Rough working

Expand the RHS
$3\sin x + 4\cos x = k\cos x \cos a - k\sin x \sin a$
Equate coefficients:
$\sin x$: $3 = -k\sin a \Rightarrow k\sin a = -3$
$\cos x$: $4 = k\cos a$

Choose option C

Obviously for the objective questions no working is required so, unlike the extended response questions covering the same topic, the same development is not needed.
However, some working will help you when it comes time to check your answers.

2   What is the maximum value of $\sin x + \cos x$?

A   1
B   −1
C   2
D $\sqrt{2}$

Rough working

Use form $r\sin(x + a)$
$r\cos a = 1$ and $r\sin a = 1 \Rightarrow r^2 = 2 \Rightarrow r = \sqrt{2}$
Max is $\sqrt{2}$

Choose option D

3   In which quadrant does a lie when
$2\sin x - 3\cos x$ is expressed in the form $r\cos(x - a)$?

A   1st quadrant
B   2nd quadrant
C   3rd quadrant
D   4th quadrant

Rough working

Expand: $2 \sin x - 3 \cos x = r \cos x \cos a + r \sin x \sin a)$
Equate coefficients: $2 = r \sin a$ (positive sine: 1st or 2nd)
$-3 = r \cos a$ (negative cosine: 2nd or 3rd)
So 2nd quadrant.

Choose option B

The formality of such a question can be raised.

❹ When $3 \sin x - \cos x$ is expressed in the form $r \sin(x + a)$ where $r > 0$ and $0 \le a \le 2\pi$, in which of the following intervals does a lie?

A $\quad 0 \le a \le \frac{\pi}{2}$
B $\quad \frac{\pi}{2} \le a \le \pi$
C $\quad \pi \le a \le \frac{3\pi}{2}$
D $\quad \frac{3\pi}{2} \le a \le 2\pi$

Rough working

Expand: $3 \sin x - \cos x = r \sin x \cos a + r \cos x \sin a$
Equate coefficients: $3 = r \cos a$ (positive cosine: 1st or 4th)
$-1 = r \sin a$ (negative sine: 3rd or 4th)

So 4th quadrant

Choose option D

❺ To which of the following expressions is $\sin x - \cos x$ equal?

A $\quad \sqrt{2} \cos(x - \frac{3\pi}{4})$
B $\quad \sqrt{2} \cos(x - \frac{\pi}{4})$
C $\quad 2 \cos(x - \frac{3\pi}{4})$
D $\quad 2 \cos(x - \frac{\pi}{4})$

Rough working

Expand: $\sin x - \cos x = r \cos x \cos a + r \sin x \sin a$
Equate coefficients: $1 = r \sin a$ (positive sine: 1st or 2nd)
$-1 = r \cos a$ (negative cosine: 2nd or 3rd)

So 2nd quadrant
Also $r = \sqrt{(1 + 1)} = \sqrt{2}$

Choose option A

6 Find the minimum value of $5 \sin x - 12 \cos x + 10$.

A   10
B -13
C -3
D -1

Rough working

For any of the forms $r = \sqrt{(5^2 + 12^2)} = 13$.

Minimum value of either $f(x) = 13 \sin(x \pm a)$ or $13 \cos(x \pm a)$ is -13
Minimum value of $f(x) + 10 = -3$

Choose option C

7 $3 \cos \theta + 4 \sin \theta$ is expressed in the form $5 \cos(\theta + a)$.
What are the values of $\sin a$ and $\cos a$ respectively.

A   0·8 and 0·6
B $-0·8$ and 0·6
C   0·8 and $-0·6$
D $-0·8$ and $-0·6$

Rough working

$3 \cos \theta + 4 \sin \theta = 5 \cos(\theta + a)$
Expanding RHS: $3 \cos \theta + 4 \sin \theta = 5 \cos \theta \cos a - 5 \sin \theta \sin a$
Equate coefficients: $4 = -5 \sin a \Rightarrow \sin a = -0·8$
$3 = 5 \cos a \Rightarrow \cos a = 0·6$

Choose option B

8 For what value of θ does the function $f(\theta) = \sin \theta + \cos \theta$ take its minimum value when $0 \le \theta \le 2\pi$?

A $\frac{5\pi}{4}$
B $\frac{3\pi}{2}$
C $\frac{\pi}{4}$
D $\frac{\pi}{2}$

Rough working

$\sin \theta + \cos \theta = r \sin(\theta + a) = r \sin \theta \cos a + r \cos \theta \sin a$
$r = \sqrt{(1 + 1)} = \sqrt{2}$
$r \cos a = 1$; $r \sin a = 1 \Rightarrow \tan a = 1 \Rightarrow a = \frac{\pi}{4}$

[Check sine positive; check cosine positive $\Rightarrow$ 1st quadrant]

Minimum occurs when $\sin(\theta + \frac{7}{4}) = -1$
$\Rightarrow \theta + \frac{7}{4} = \frac{3\pi}{2}$
$\Rightarrow \theta = \frac{3\pi}{2} - \frac{7}{4} = \frac{5\pi}{4}$

Choose option A

This is an A/B grade question.

A basic short response question that needs a calculator

❾ Express $5\cos x° - 12\sin x°$ in the form $k\sin(x + a)°$
where $k > 0$ and $0 < a < 360$.

⟮Response⟯────────────────────────────────

$5\cos x° - 12\sin x° = k\sin(x + a)°$
$\Rightarrow 5\cos x° - 12\sin x° = k\sin x° \cos a° + k\cos x° \sin a°$
Equating coefficients:
$5 = k\sin a°$ ... (1st or 2nd quadrant)
$-12 = k\cos a°$ ... (2nd or 3rd quadrant)
$\Rightarrow$ 2nd quadrant.
$\tan a° = -\frac{5}{12}$
$\Rightarrow a = -22\cdot6$ or $-22\cdot6 + 180 = 157\cdot4$
Since a is in second quadrant then $a = 157\cdot4$
$k^2\sin^2 a° + k^2\cos^2 a° = 5^2 + (-12)^2$
$\Rightarrow k^2(\sin^2 a° + \cos^2 a°) = 169$
$\Rightarrow k^2 = 169$
$\Rightarrow k = 13$
So $5\cos x° - 12\sin x° = 13\sin(x + 157\cdot4)°$. ────────⬭

Marking scheme

- ●^{1} expand $k\sin(x + a)°$ ... **stated explicitly**   *ss*
- ●^{2} compare coefficients ... again be explicit   *ic*
- ●^{3} evaluate k *pd*
- ●^{4} evaluate a and quote answer   *pd*

Although you may find it possible with these problems to state k and a and hence write down the answer using only your calculator ... to do so will mean that marks 1 and 2 **will be lost**.

An extended response question that needs a calculator

10 The function $f(x) = 6 \sin x + \cos x$.

(a) Show that $f(x)$ can be expressed in the form $k \sin(x - a)$ where $k > 0$ and $0 \le a \le 2\pi$, stating the values of a and k. *4 marks*

(b) Working in the interval $0 \le x \le 2\pi$, find the maximum and minimum values of the function and find the values of x at which they occur. *4 marks*

(*Response*)

(a)
$6 \sin x + \cos x = k \sin(x - a)$
$\Rightarrow 6 \sin x + \cos x = k \sin x \cos a - k \cos x \sin a$
Equating coefficients:
$6 = k \cos a$ ... (1st or 4th quadrant)
$1 = -k \sin a$ ... (3rd or 4th quadrant)
$\Rightarrow$ 4th quadrant.
$\tan a = -\frac{1}{6}$ [Make sure your calculator is set to radians.]

> **!**
> 1 degree = 0·01745 radians approx.
> If you only work to 1 decimal place, you could be making an error of ± 0·05 radians.
> 0·05 radians = 2·9° ... a big error to make. Work to 3 decimal places.

$\Rightarrow a = -0·165$ or $-0·165 + 2\pi = 6·118$
Since a is in 4th quadrant then $a = 6·118$
$k^2 \sin^2 a + k^2 \cos^2 a = 1^2 + (-6)^2$
$\Rightarrow k^2 (\sin^2 a + \cos^2 a) = 37$
$\Rightarrow k^2 = 37$
$\Rightarrow k = \sqrt{37}$
So $6 \sin x + \cos x = \sqrt{37} \sin(x - 6·118)$

(b)
Maximum value $= \sqrt{37}$ which will occur when $\sin(x - 6·118) = 1$
$\Rightarrow x - 6·118 = -\frac{3\pi}{2}, \frac{\pi}{2}, ...$ (notice that we will be adding 6·118 to our values so we should consider negative values)
$\Rightarrow x = 1·406, 7·689$
$\Rightarrow x = 1·406$ in the range.
Minimum value $= -\sqrt{37}$ which will occur when $\sin(x - 6·118) = -1$
$\Rightarrow x - 6·118 = -\pi/2, 3\pi/2, ...$ (again we should consider negative values)
$\Rightarrow x = 4·547, 10·830$
$\Rightarrow x = 4·547$ in the range.

Marking scheme

- $\bullet^1$ expand $k \sin(x - a)$ ... **stated explicitly**   *ss*
- $\bullet^2$ compare coefficients ... **again be explicit**   *ic*
- $\bullet^3$ evaluate k *pd*
- $\bullet^4$ evaluate a *pd*
- $\bullet^5$ use form $\sqrt{37} \sin(x - 6{\cdot}118)$ *ic*
- $\bullet^6$ find maximum and minimum   *pd*
- $\bullet^7$ at max $x = 1{\cdot}406$ ... and no further answer   *pd*
- $\bullet^8$ at min $x = 4{\cdot}547$ ... and no further answer   *pd*

Notes:

(a)   A mark is deducted for a student who works in degrees.

(b)   If mark 7 is lost by not considering the negative possibilities then mark 8 will not be lost for the same error.

(c)   Marks 1 and 2 go missing for the candidate who does not show his/her steps.

(d)   Finding the values of x at which extrema occur is an A/B grade skill.

An extended response question that doesn't need a calculator

⓫   (a)   Express $\sqrt{3} \sin x - \cos x$ in the form $k \sin(x - a)$
          where $k > 0$ and $0 \leq a \leq 2\pi$.   *4 marks*

   (b)   The sketch shows the graphs of $y = \sqrt{3} \sin x$ and $y = 1 + \cos x$
          crossing at two places in the domain $0 \leq x \leq 2\pi$.   *5 marks*

Find the co-ordinates of both points.

(Response)

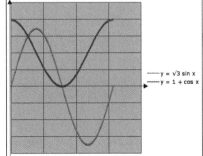

(a) $\sqrt{3} \sin x - \cos x = k \sin(x - a)$
$\Rightarrow \sqrt{3} \sin x - \cos x =$
$k \sin x \cos a - k \cos x \sin a$
Equating coefficients:
$\sqrt{3} = k \cos a$ ... (1st or 4th quadrant)
$-1 = -k \sin a$ ... (1st or 2nd quadrant)
$\Rightarrow$ 1st quadrant.

$\tan a = \frac{1}{\sqrt{3}}$

$\Rightarrow a = \frac{\pi}{6}$ (being in 1st quadrant)

$k^2 \sin^2 a + k^2 \cos^2 a = 1^2 + (\sqrt{3})^2$

$\Rightarrow k^2 (\sin^2 a + \cos^2 a) = 4$

$\Rightarrow k^2 = 4$

$\Rightarrow k = 2$

$\sqrt{3} \sin x - \cos x = 2 \sin(x - \frac{\pi}{6})$

(b)

$y = \sqrt{3} \sin x$ and $y = 1 + \cos x$ cross when

$\sqrt{3} \sin x = 1 + \cos x$

$\Rightarrow \sqrt{3} \sin x - \cos x = 1$

$\Rightarrow 2 \sin(x - \frac{\pi}{6}) = 1$

$\Rightarrow \sin(x - \frac{\pi}{6}) = \frac{1}{2}$

$\Rightarrow x - \frac{\pi}{6} = \frac{\pi}{6}, \pi - \frac{\pi}{6} = \frac{5\pi}{6}$

$\Rightarrow x = \frac{\pi}{3}, \pi$

$\Rightarrow y = 1 + \cos \frac{\pi}{3} = \frac{3}{2}, 1 + \cos \pi = 0$

The points are $(\frac{\pi}{3}, \frac{3}{2})$ and $(\pi, 0)$.

Marking scheme

- $\bullet^1$ expand $k \sin(x - a)$ ... **stated explicitly**   *ss*
- $\bullet^2$ compare coefficients ... again be explicit   *ic*
- $\bullet^3$ evaluate k *pd*
- $\bullet^4$ evaluate a *pd*
- $\bullet^5$ form an equation   *ss*
- $\bullet^6$ translate info into another suitable form
  [use form $2 \sin(x - \frac{\pi}{6})$ in equation]   *ic*
- $\bullet^7$ solve for $x = \frac{\pi}{3}, \pi$ *pd*
- $\bullet^8$ find corresponding y values   *pd*
- $\bullet^9$ Express solution in language appropriate to situation
  [the points are $(\frac{\pi}{3}, \frac{3}{2})$ and $(\pi, 0)$]   *ic*

Notes:

(i)   Marks 7 and 8 can be 'cross-marked'. This means that you might gain mark 7 for one pair of corresponding values of x and y.
This allows the candidate who only answers $x = \frac{\pi}{3}$ and $y = \frac{3}{2}$ after all their work to gain one mark at least.

(ii)   It is possible to look at the diagram and guess the solutions.
The examiner will never knowingly use a misleading diagram.
However, as you can imagine, no credit will be given if you guess!

Continued on next page

Answers obtained from scale drawings are not worthy of a mark at Higher.

(iii) Note how part (a) provides C-grade candidates with an 'in' to the problem. It also points to the strategy for the rest of the question. It is perfectly possible that the same question could have been asked without such a lead in but most candidates would then have found it inaccessible.

Imagine if the question had been:

Solve the system of equations
$y = \sqrt{3} \sin x$
$y = 1 + \cos x$
for $0 \le x \le 2\pi$

(iv) The expression '...in the form $k \sin(x - a)$' also acts as a prompt to use this strategy. However, don't depend on that phrase, or something like it, always being there.

There have been times when no guidance has been given and the question simply took the form:

'Solve the equation $2 \sin x + 5 \cos x = 4$.'

This wording leaves the candidate to select a strategy.

An extended response question that needs a calculator

17 (a) Find the minimum value of $f(x) = \dfrac{1}{\sin x + \cos x}$, $0 \le x < \tfrac{3\pi}{4}$ and the value of x at which it occurs. *7 marks*

(b) Comment on the upper bound of the domain of the function. *1 mark*

(*Response*)

(a) The minimum value of a fraction will occur when the denominator is at its maximum. So we must find the maximum value of $\sin x + \cos x$.

$\sin x + \cos x = k \sin(x + a)$

$\Rightarrow \sin x + \cos x = k \sin x \cos a + k \cos x \sin a$

Equating coefficients:

$k \cos a = 1$ (1st or 4th quadrants) and $k \sin a = 1$ (1st or 2nd quadrants)

$\tan a = 1 \Rightarrow a = \tfrac{\pi}{4}$ (1st quadrant)

$k^2 \sin^2 a + k^2 \cos^2 a = 1^2 + 1^2 = 2$

$\Rightarrow k^2 (\sin^2 a + \cos^2 a) = 2$

$\Rightarrow k^2 = 2$

$\Rightarrow k = \sqrt{2}$

$\Rightarrow \sin x + \cos x = \sqrt{2} \sin(x + \frac{\pi}{4})$

This is at a maximum of $\sqrt{2}$ when $\sin(x + \frac{\pi}{4}) = 1$

$\Rightarrow x + \frac{\pi}{4} = \frac{\pi}{2}$

$\Rightarrow x = \frac{\pi}{4}$

So the minimum value of $f(x) = f(\pi/4) = 1/\sqrt{2}$.

(b) When $x = \frac{3\pi}{4}$, $\sin x + \cos x = 0$ and $\frac{1}{0}$ is undefined. ⎯⎯⎯⎯⎯⎯◯

Marking scheme

- $\bullet^1$ expand $k \sin(x + a)$ ... **stated explicitly**   ss
  [this could also be $k \sin(x - a)$ or $k \cos(x \pm a)$]
- $\bullet^2$ compare coefficients ... again be explicit   ic
- $\bullet^3$ evaluate k pd
- $\bullet^4$ evaluate a pd
- $\bullet^5$ communicate conditions for minimum   ic
- $\bullet^6$ find minimum for $f(x)$ pd
- $\bullet^7$ state corresponding x pd
- $\bullet^8$ communicate reason for open bound   ic

Topic Tips

As you will have noticed this particular unit of work is quite singular in its aims.

It is a relatively straightforward process and should be considered wherever you spot sums and differences of sines and cosines.

Practise the skill and consider maxima and minima.

Remember that you exhibit an A/B skill when you find the angle corresponding to these extrema.

Be sensitive to the use of radians and degrees. Marks can be easily lost.

Finally, two marks out of the four will go missing if you do not show the expansion and the subsequent equating of coefficients.

The SQA have stated the following:

Candidates should be encouraged to show all intermediate working, e.g. the calculation of r and α, the expansion of trigonometric formulae and the equating of coefficients.

16 | Problem Solving

> *Interpretation*
>
> *Selection of strategy*
>
> *Implementation of strategy*
>
> *Communication*

The main aim of the Higher Mathematics course is to develop problem-solving skills. The main aim of the assessment is to measure your ability to cope with problems. In their conditions and arrangements document, the SQA offer advice and guidance as to what this might entail and how to distinguish between C-level and A/B-level candidates.

For each problem there are considered to be four steps: interpretation, selection of strategy, implementation of strategy, and communication.

Interpretation

Read the question and decide what is relevant.
The data can be both qualitative and quantitative and can appear in
(i) real-life scenarios,
(ii) material from other subjects,
(iii) other familiar areas of mathematics.

'Grade A performance is demonstrated through coping with the interpretation of more complex contexts requiring a higher degree of reasoning ability.'

Selection of strategy

Decide on what steps you will take to solve the problem.
Make the steps plain for the examiners to see. Don't assume they'll know where you got your answers. They will be awarding marks for demonstrating steps when selecting an algorithm, or sequence of algorithms.

'Grade A performance is demonstrated by using a more extended sequence of steps or applying a more complex strategy or applying it to a more complex context.'

Implementation of strategy

You should be able to use what knowledge and skills you have learned to enact your strategy and to arrive at a result by:
(i)   processing data ... both numerical and symbolic;
(ii) marshalling the facts and sustaining logical reasoning;
(iii) sustaining and exhibiting logical reasoning;
(iv) appreciating the need for justification and proof.

Again the examiner should not be left guessing what you are doing – your working and motives should be on the page.

'Grade A performance is demonstrated by being able to process data in more complex situations, and sustaining logical reasoning, where the situation is more removed from the maths you will need to apply.'

Communication

You will have to decide on the most appropriate way to report your answer. This should be done with regard to the context of the problem. Candidates can lose marks when, having done all the work, they don't answer the question that was actually asked.
You should express the solution in a language that suits the situation being mindful of the use of units and such like. Form is also important, e.g. if asked for the coordinates of a point, then finding x and then five lines later finding y, may not be enough. Pull your results together and give your findings in the form (x, y).

'Grade A performance is demonstrated by being able to communicate in more complex situations and in unfamiliar contexts.'

To help you practice further, please see the complete practice Higher Maths exam on the Leckie & Leckie website (*www.leckieandleckie.co.uk*). This also has a complete marking scheme and worked answers to help you with your practice.

We also recommend you obtain a copy of Leckie & Leckie's Official SQA Past Papers in Higher Maths. These contain the 5 most recent Higher Maths exams, along with the answers to all these exams' questions.

All that remains is to practice well before the final test:
interpret data, select strategies, process data, communicate answers ... as expressively as you can on paper. **Marks make grades ... and you make your own luck**.